Nelson Maths

4

Pupil Book

Karen Morrison
Lisa Greenstein

OXFORD
UNIVERSITY PRESS

OXFORD
UNIVERSITY PRESS

Great Clarendon Street, Oxford, OX2 6DP, United Kingdom

Oxford University Press is a department of the University of Oxford.

It furthers the University's objective of excellence in research, scholarship, and education by publishing worldwide. Oxford is a registered trade mark of Oxford University Press in the UK and in certain other countries.

British Library Cataloguing in Publication Data

Data available

ISBN: 978-1-382-01004-7

1 3 5 7 9 10 8 6 4 2

Paper used in the production of this book is a natural, recyclable product made from wood grown in sustainable forests. The manufacturing process conforms to the environmental regulations of the country of origin.

Printed in Great Britain by Bell and Bain Ltd, Glasgow

Acknowledgements

The publisher and authors would like to thank the following for permission to use photographs and other copyright material:

Cover: Matthieu Nivesse. **Photos: p12:** James Clarke/Shutterstock; **p18(l):** Celiafoto/Shutterstock; **p18(r):** Sheila Fitzgerald/Shutterstock; **p19:** Charlie Riedel/AP/Shutterstock; **p27(l):** Andrey Mertsalov/Shutterstock; **p27(r):** Mega Pixel/Shutterstock; **p45(tr):** mountainpix/Shutterstock; **p45(tl):** DG FotoStock/Shutterstock; **p45(mr):** Dimitris Leonidas/Shutterstock; **p45(bl):** Oksana Shufrych/Shutterstock; **p64:** Cesare Palma/Shutterstock; **p65:** ericsphotography/iStockphoto; **p70:** M. Unal Ozmen/Shutterstock; **p81:** Gregory Gerber/Shutterstock; **p129:** Menno Schaefer/Shutterstock; **p134:** dpa picture alliance /Alamy Stock Photo.

Artwork by Aviel Basil, Q2A Media, Alan Rogers, Pantek Media, and OKS Prepress.

Every effort has been made to contact copyright holders of material reproduced in this book. Any omissions will be rectified in subsequent printings if notice is given to the publisher.

Contents

Think maths

What do you think?

 Think and share

Zara drew this pattern for a cube.

Which of these cubes could be made from Zara's pattern?

a  b 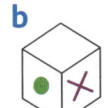 c  d

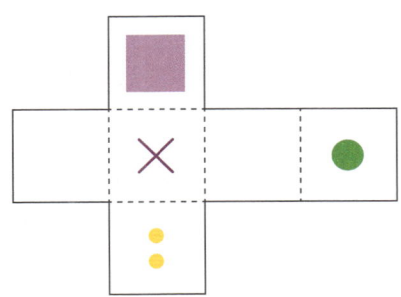

1 Think about the activity and talk about these questions in your group.

a What did you have to think about to solve the problem?

b What did you try to help you solve the problem?

c What was the most challenging part of this activity?

d Did you ever feel you didn't know what to do? What did you try to move past that?

e Did you make any mistakes? How did these help you learn?

2 Choose three of these statements about learning maths.

Everyone can learn maths.

Mistakes are valuable for learning maths.

Questions are really important.

Maths is about being creative and making sense of things.

Maths class is a place to learn, not a place to perform.

Understanding things properly is more important than doing things quickly.

Take turns with your partner. Try to persuade each other that the statements you chose are true.

➡ *Workbook page 5*

Maths helps your brain grow

Your brain grows and makes connections when you think hard. When you make mistakes, your brain grows even more. When you try hard without giving up, you learn more. Your brain makes new connections and grows stronger.

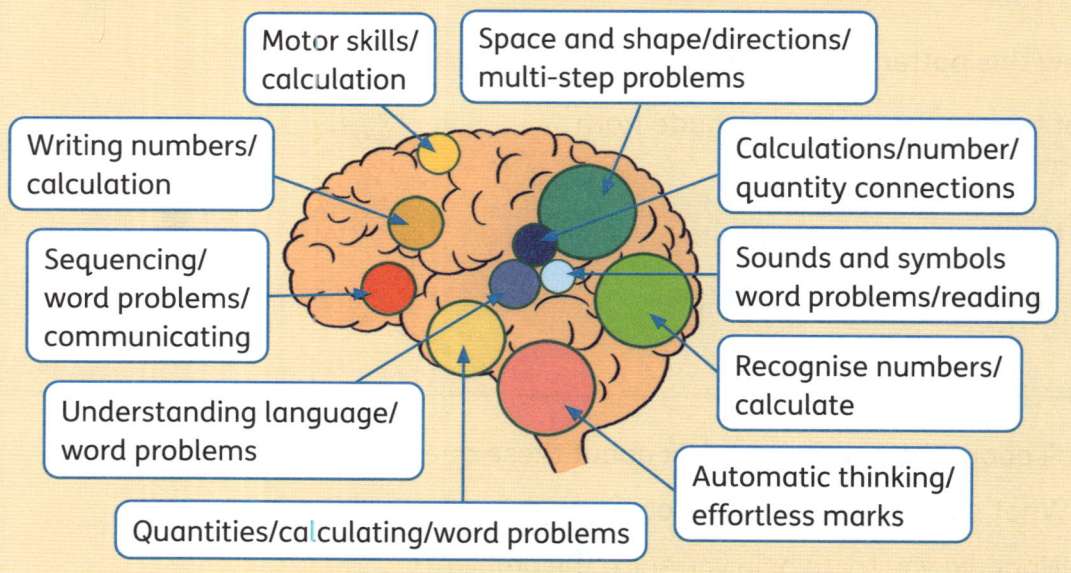

Motor skills/calculation

Space and shape/directions/multi-step problems

Writing numbers/calculation

Calculations/number/quantity connections

Sequencing/word problems/communicating

Sounds and symbols word problems/reading

Understanding language/word problems

Recognise numbers/calculate

Quantities/calculating/word problems

Automatic thinking/effortless marks

This drawing shows the areas of your brain you use when you do different kinds of maths.

What maths problem would use at least four areas of your brain?

Problem solving

1 Ali, Jade and Zoe coloured triangles and folded them up to make paper planes.

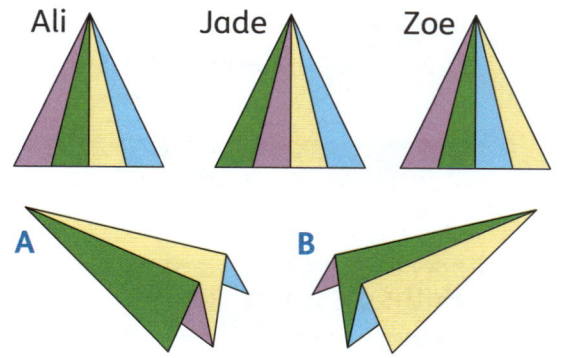

Ali   Jade   Zoe

A   B

Here are two folded planes.

a Can you work out which plane belongs to which pupil?

b Use tracing paper to trace the outline of a folded plane (from part a). Then, for the plane that is *not* shown in part a, colour your outline to show what that plane looks like.

➡ *Workbook page 6*

Thinking about shapes

A **net** is a flat pattern that you can fold up to make a model of a (3D) shape.

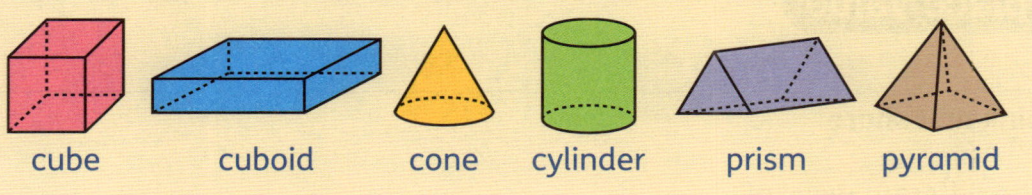

cube cuboid cone cylinder prism pyramid

1 One face is missing from each of these nets of the 3D shapes shown above.

- Work out which face is missing.
- Where would it go on the net?

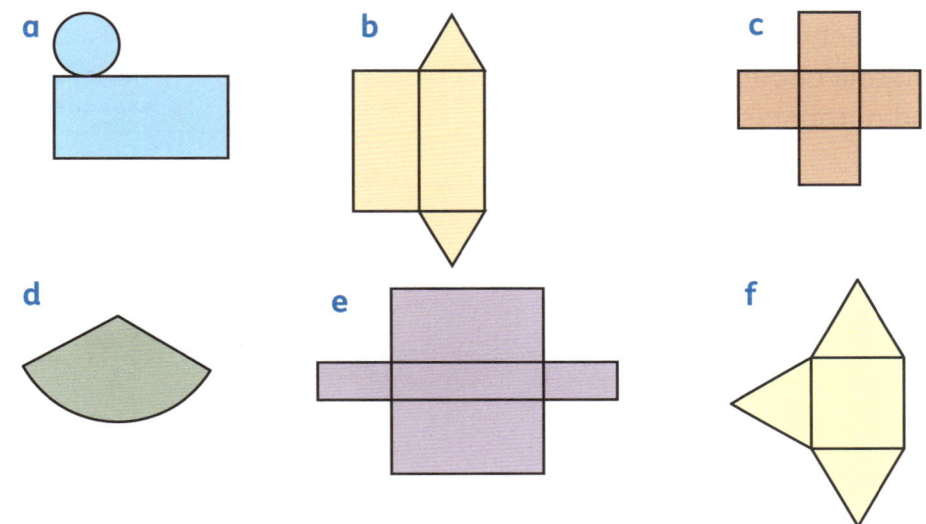

a   b   c

d   e   f

2 Which of the nets below will *not* make a cube or a cuboid? Give a reason for each choice.

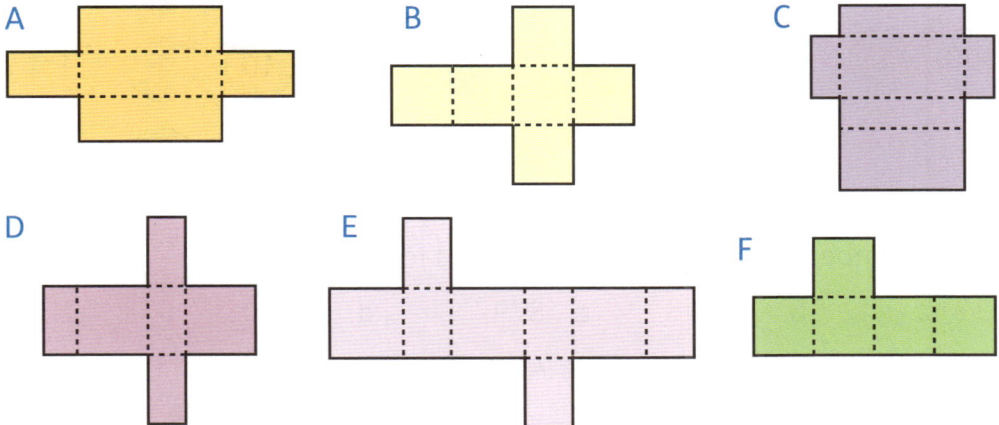

A   B   C

D   E   F

Number and place value

Revisit place value

 Think and share

Look at the different ways of representing a number.

*   What is the difference between the **place** and the **value** of a **digit**?

*   Why is it important to use the correct words when you talk about digits and numbers?

Place is hundreds (H)

Value is 600 → **625** 625 is a number. We say: six hundred and twenty-five

6, 2 and 5 are digits. This is a 3-digit number.

Place-value table

Hundreds	Tens	Ones
6	2	5

$625 = (6 \times 100) + (2 \times 10) + (5 \times 1)$
$= 600 + 20 + 5$

This is called **expanded form**.

1 Count and work out the number. Write each number using expanded form.

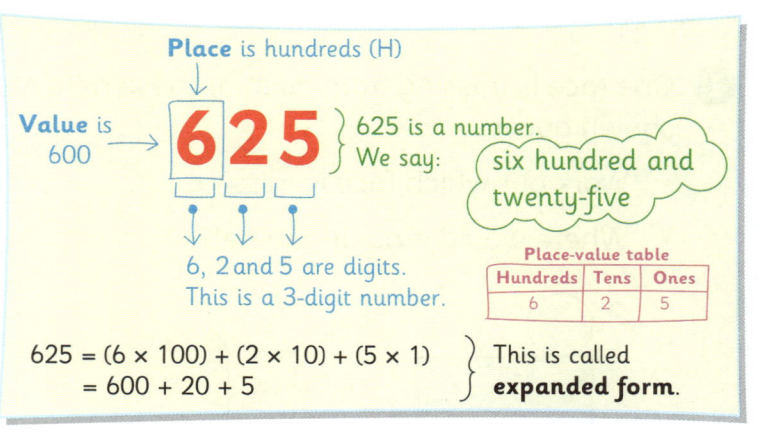

a How many bags of crisps altogether?

b How many mints altogether?

2 Write the totals shown by these place-value counters.

a (100) (1) (1) (1) (1) (100) (1) (1) (1)

b (100) (100) (10) (1) (100) (10) (10) (1) (1) (100) (10) (10) (1)

3 Say each number. Then write it in expanded form.

a 613   b 226   c 879   d 547

e 984   f 732   g 461   h 358

➡ *Workbook page 7*

Place value to thousands

The greatest 3-digit number we can make is 999. For numbers greater than 999, we need another place to write 4-digit numbers.

10 ones make 1 ten.
10 tens make 1 hundred.
10 hundreds make 1 thousand. We write this as 1000.
Two thousand, three hundred and forty-nine is written as 2349.
We extend the place-value table to include thousands like this:

Thousands	Hundreds	Tens	Ones
2	3	4	9

1 Say each number.

a 2812　　b 9322　　c 6871

d 3562　　e 8447　　f 4924

2 Write these numbers using digits.

a Five thousand, seven hundred and ninety-two

b Eight thousand, two hundred and seventy-five

c Three thousand, six hundred and fifty

d One thousand, nine hundred and sixty

e Two thousand, four hundred and eighty-nine

f Nine thousand, seven hundred and sixty-five

g Four thousand, two hundred and fifty-seven

h Six thousand, two hundred and eighty-three

3 What is the value of the underlined red digit in each of these numbers?

a 3<u>6</u>12　　b 948<u>6</u>　　c 20<u>3</u>2

d <u>3</u>009　　e 36<u>2</u>0　　f 4<u>9</u>15

g <u>5</u>106　　h 99<u>9</u>9　　i 135<u>9</u>

4 Describe the relationship between 3500, 350 and 35.
You can draw diagrams to show what you mean.

Workbook page 8

Expanded form

You can **partition** and expand 4-digit numbers.

Partition the number 4765.

4765 = 4 thousands, 7 hundreds, 6 tens and 5 ones

Write the number 4765 in expanded form.

$4765 = (4 \times 1000) + (7 \times 100) + (6 \times 10) + (5 \times 1)$
$\quad\quad = \quad 4000 \quad + \quad 700 \quad + \quad 60 \quad + \quad 5$

1 Write each number in expanded form.

a	5240	**b**	1098	**c**	1609	**d**	3182
e	8056	**f**	7484	**g**	6179	**h**	9762

2 Write these numbers. Say each one.

a 4000 + 300 + 30 + 5   **b** 8000 + 500 + 20 + 9

c 4000 + 300 + 20 + 1   **d** 5000 + 400 + 80 + 9

e 3000 + 200 + 30   **f** 2000 + 40 + 9

g 9000 + 50   **h** 7000 + 500 + 3

3 Copy and complete these number sentences.

a 1876 = 1000 + ☐ + 70 + 6   **b** 3876 = ☐ + 800 + 70 + 6

c 4265 = 4000 + 200 + 60 + ☐   **d** 9398 = ☐ + 300 + ☐ + 8

e 3098 = ☐ + 90 + 8   **f** 5008 = ☐ + 8

💡 **Problem solving**

Make a table or organised list.

4 Find a 4-digit whole number that:

- is between 2000 and 3000
- is an even number
- can be divided by 5
- is a **multiple** of 9.

➡ *Workbook page 9*

Compare and order numbers

A **number line** is a straight line that shows numbers equally spaced along its length.

The number line below goes from 0 to 1000.
Not all numbers are marked on the number line.
Each **vertical** mark (I) represents a jump of one hundred.
Numbers go up in value as you move to the right along the number line.

The blue arrow shows the position of 300.

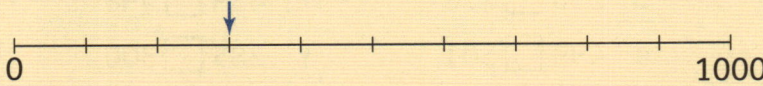

You can **estimate** the position of different numbers on a number line.
For example, 350 is halfway between 300 and 400.

1 For each number line, write down the numbers that match each letter.

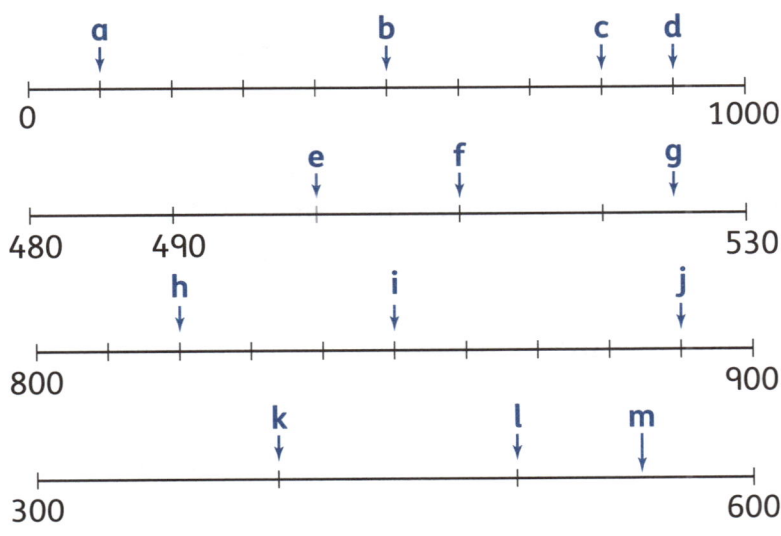

2 This is a blank 0–10 000 number line.

  a Estimate the value of each number marked with an arrow.

  b Tell your partner how you decided what each number is.

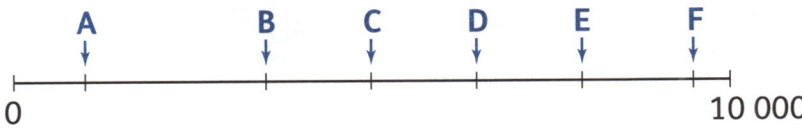

➡ *Workbook page 10*

Use < and > to compare and order numbers

< means 'less than' $\boxed{100 < 105}$

> means 'greater than' $\boxed{105 > 100}$

You compare numbers by deciding which one is greater in value.

1 Copy and complete. Fill in < or >.

a 199 ☐ 150   b 210 ☐ 270   c 439 ☐ 498

d 760 ☐ 719   e 499 ☐ 501   f 285 ☐ 300

2 Use these number cards. Write:

a three number sentences using <

b three number sentences using >.

 145  100  638  249  975  728

3 Write each set of numbers in **order** from smallest to greatest.

a 720, 659, 985, 460, 302   b 118, 102, 108, 111, 150

c 907, 970, 709, 790, 799, 797   d 832, 328, 382, 823, 238

4 Say whether each statement is true or false.

a 455 > 400 + 50   b 799 < 797

c 309 < 3 × 100 and 9 × 1   d 588 < 5 × 100 and 80 + 8

 **Problem solving**

Use a number line.

5 Four pupils tried to guess how many sweets were in a jar.

These are their guesses:

432   409   411   398

The actual number of sweets was 422. Which guess was closest?

➡ *Workbook page 11*

Compare and order larger numbers

1 Write down the greatest number in each set.

   a 1326, 1632, 1623   **b** 4779, 4797, 4977

   c 2581, 2851, 2815   **d** 6239, 6329, 6392

2 Look again at the greatest number in each set in question 1. Write the number that is:

   a ten smaller than that number

   b one hundred greater than that number

   c two thousand greater than that number.

3 Write each set of numbers in order from smallest to greatest.

a	6125	1372	5827	3150	6324
b	2895	3199	1596	2250	1826
c	5643	4291	6126	5242	4871
d	9875	8982	9645	8777	7958
e	6708	7729	5936	6599	6821
f	1986	2135	1520	2448	1992

 Problem solving

> There may be more than one correct answer.

4 Rewrite each number sentence. Write digits in place of the * to make each number sentence true.

   a 2460 > 2*60   **b** 4**9 < 4119

   c 5099 < *999   **d** 4819 < 4*20

   e 6568 < 6*6*   **f** *312 > 8547

5 Find the number that is halfway between the two numbers in each number sentence you wrote in question 4. Draw rough number lines to help you if you need them.

→ Workbook page 12

Round numbers to the nearest 10

It took Naresh 24 minutes to walk home.

It took Kamala 27 minutes to walk home.

- Who took about 30 minutes to walk home? How did you decide?

A **rounded number** is an estimated value that is close to the real number. When we **round** a number to the nearest 10, we work out which ten it is closest to.

Look at the number line to see how to round 24 and 27 to the nearest 10.

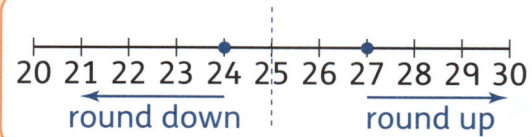

- 24 is between 20 and 30, but it is closer to 20, so 24 rounds to 20.
- 27 is also between 20 and 30, but it is closer to 30, so 27 rounds to 30.

To round to the nearest 10, look at the tens and ones.

- For 1, 2, 3 or 4 ones, leave the tens digit the same and write 0 in the ones place.
- For 5, 6, 7, 8 or 9 ones, add 1 to the tens digit and write 0 in the ones place.

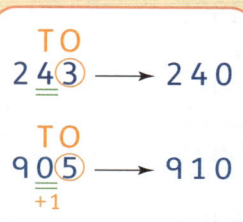

All multiples of 10 have 0 in the ones place. So, when you round to the nearest 10, the rounded number will have 0 ones.

1 Round each number to the nearest 10.

    a 148   **b** 652   **c** 142   **d** 1405

2 Give an example of why you might round each of these:

    a prices   **b** **heights**   **c** **mass**   **d** times

3 Read the statements. Do you agree or disagree? Why?

a

> I want to fence a **perimeter** of 84 **metres**. The fencing comes in 10 metre rolls.
>
> I can round 84 metres to 80 and buy 8 rolls.

b

> The correct dose of a medicine is 125 **millilitres** every 6 hours.
>
> That's hard to measure so I can round it to 130 millilitres.

➡ *Workbook page 13*

Round numbers to the nearest 100

To round to the nearest 100, look at the digit in the tens place.

- If that digit is 1, 2, 3 or 4, leave the hundreds unchanged and write zeros in the tens and ones places.
- If the digit in the tens place is 5, 6, 7, 8 or 9, add 1 to the hundreds digit and write zeros in the tens and ones places.

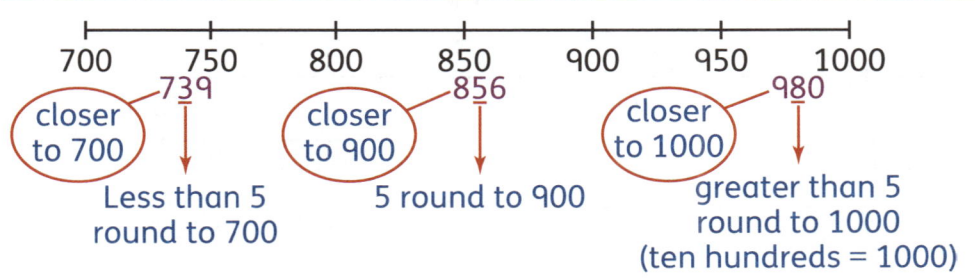

Round 1439 to the nearest 100.

1439 The digit in the tens place is 3, so the number rounds to 1400.

Round 3689 to the nearest 100.

3689 The digit in the tens place is 8, so round to the next 100, which is 3700.

1 Round each number to the nearest 100.

a 417	b 475	c 450	d 4265
e 4851	f 4359	g 4672	h 4499

2 Round each number to the nearest 100.

a 2449	b 6921	c 8689	d 3162
e 5863	f 7500	g 4425	h 9267
i 3318	j 1282	k 4099	l 8989

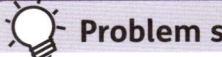 **Problem solving**

Draw a number line to help you.

3 Each of these numbers was rounded to the nearest 100. What is the smallest and the greatest number each could have been to start with?

a 2300	b 5600	c 3100	d 7900

➡ *Workbook page 13*

Round numbers to the nearest 1000

To round a number to the nearest 1000, use the same *method* you used to round to the nearest 100 and 10. Use the digit in the hundreds place to decide what happens to the thousands.

Round 4215 and 4652 to the nearest 1000.

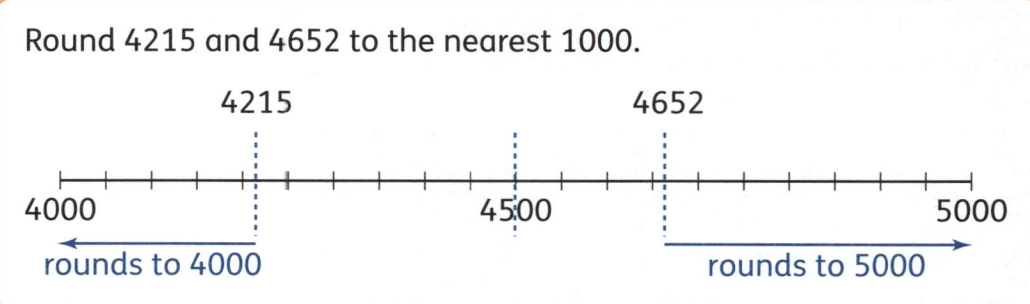

1   Say whether each number would round to 4000 or 5000 to the nearest 1000.

    a 4170   b 4753   c 4500   d 4265

    e 4851   f 4359   g 4672   h 4499

    i 4900   j 4820   k 4065   l 4577

2   Round each number to the nearest thousand.

    a 2449   b 6921   c 8689   d 3162

    e 5863   f 7500   g 4425   h 9267

    i 3318   j 1282   k 4099   l 8989

Think back to what you did on page 15.

 Problem solving

3   These numbers have been rounded to the nearest thousand. What is the smallest, and greatest, number they could have been?

    a 2000   b 5000   c 3000   d 7000

4   What patterns do you notice in your answers to question 3?

5   How can these patterns help you answer similar questions?

➡ *Workbook page 14*

Work with rounded numbers

Rounded numbers are useful for estimating.

- You can round numbers to help you do calculations in your head. This helps you check that your answer to a calculation is about right.
- Rounded amounts can help you make decisions. You can use rounded prices to see whether you have enough money when you are shopping.
- Newspapers and news websites often use rounded numbers in headlines. For example: 'One third of the 8000 known species of amphibians are endangered'.

How would you round these figures for use in news headlines?

a   Average price of hand sanitiser rises to £99 per case.

b   Workers to receive a bonus of £399 per month.

c   Average rainfall for this month is a record 1298 **millimetres**.

d   7231 people register for online service in first hour.

e   Tsunami waves travel at speeds of 803 km/hr.

1 Naresh says '3540 rounded to the nearest hundred is 3600 because when there is a 5 in the place you are rounding to, it goes up by 1'.

   a   What interesting mistake has Naresh made?

   b   Draw a number line to help Naresh reach the correct answer.

2 Estimate each of the following amounts. The first one has been done for you.

   a   39 + 13 *This is approximately* 40 + 10 = 50

   b   37 + 65   c   53 + 126   d   23 + 61 + 59

   e   49 – 23   f   437 – 19   g   312 – 46

3 In a group of 4297 children, just over half of them are under the age of 5. Approximately how many children is that?

4 3982 people register for a marathon. Each person who enters gets a T-shirt. Runners who finish in less than 6 hours get a medal.

   a   The organisers order 4000 T-shirts and 3900 medals. How do you think they decided on these numbers?

   b   Why do you think they ordered more T-shirts than the number of runners, but fewer medals than runners?

Roman numerals

Have you ever seen numbers like these on buildings or clocks?

The numbers are called **Roman numerals**. They are written using combinations of the letters I, V, X, L, C, D and M. Each letter represents a number.

I	V	X	L	C	D	M
1	5	10	50	100	500	1000

The Roman number system does not use **place value** and it has no zero (0). You combine letters using a set of rules to make different numbers.

Rule	Example
Letters cannot be written more than three times in a row.	I = 1 II = 2 III = 3
If the letters are the same, and next to each other, you add them to find the number.	XXX = 10 + 10 + 10 = 30
If letters of a smaller value are written to the *right* of letters of a higher value, you add them too.	XXV = 10 + 10 + 5 = 25
If letters of a smaller value are written to the *left* of letters of a greater value, you **subtract** them.	IX = 10 − 1 = 9 XL = 50 − 10 = 40
If a letter of smaller value is between two letters of greater value, you subtract it from the one on the right.	XIV = 10 + (5 − 1) = 10 + 4 = 14

1 Write the answers to these questions using Roman numerals.

   a How old are you?

   b How many people live in your home?

   c How many players are there in a football (soccer) team?

   d What grade are you in at school this **year**?

   e What number is shown on the building in the photograph?

   f How is 12 shown on the clock face in the photograph?

➡ *Workbook page 15*

More Roman numerals

1 Write the numbers that match each of these Roman numerals.

   a XIV   **b** XII   **c** LX

   d LV   **e** XL   **f** LIX

2 Look at this photo from a sporting competition that takes place every year.

The Roman numerals tell you the number of times the competition has taken place.

   a How many times had the competition taken place when this photo was taken?

   b Write the Roman numerals for the previous year.

   c Design a logo with the year in Roman numerals for next year's competition.

3 Count in 5s from 50 to 100. Write the numbers you count using Roman numerals.

4 Read what a pupil said below. Do you agree or disagree? Explain why.

> In our number system, multiples of 10 all have zero in the ones place. The Roman numeral for 10 is X, so I think all Roman numeral multiples of 10 must have an X in them somewhere.

💡 Problem solving

5 Work out the rule for this number pattern and write the missing numbers using Roman numerals.

XVIII → XXXVIII → LVIII → ☐ → ☐

2D shapes

Revisit 2D shapes

 Think and share

What do these road signs mean? If you don't know, how could you find out?

a

b

c

d

- What shape is each sign?
- Which shape of road sign do you see most in your area?
- Why do you think this shape is used more than others?
- For each shape above, give at least one other example of where you see this shape in real life.

1 Look at these shapes, which have been made from yellow string.

- Count the number of **sides** each shape has.
- What is the mathematical name of each shape?

a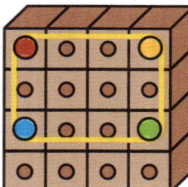

b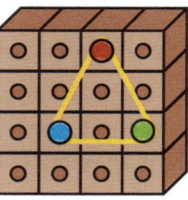

c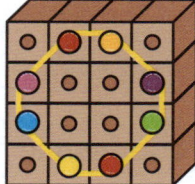

d

e

f

Shapes and their properties

A **polygon** is a 2D shape with three or more straight sides. When all the sides are the same **length** and all the **angles** are the same size, the polygon is **regular**. The sides meet at corners. We call these **vertices**.

Shape	Examples	Properties		
		Number of sides	Number of angles	Number of vertices
Triangle		3	3	3
Quadrilateral		4	4	4
Pentagon		5	5	5
Hexagon		6	6	6
Heptagon		7	7	7
Octagon		8	8	8

1 Count the number of sides on each shape.
Write the name of each shape and say whether it is a regular polygon or not.

 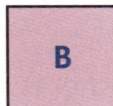 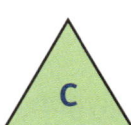 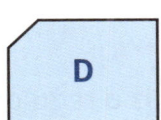 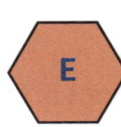 

2 Draw three polygons that have:

a 6 vertices

b 4 equal sides

c a **right angle** and three sides.

➡ *Workbook page 16 and page 17*

Shapes and angles

This is a right angle.

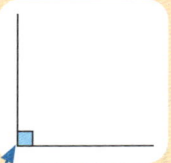

We can show a right angle like this.

This shape is a **quadrilateral**.

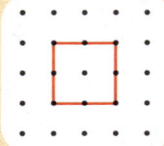

It has 4 sides.
It also has 4 right angles.

A right angle is a quarter **turn**.

1 How many right angles do each of these shapes have?

a

b

c

2 Draw these shapes on dotted paper:

a a quadrilateral with only 1 right angle

b a pentagon with 3 right angles

c a hexagon with 2 right angles

d a quadrilateral with 2 right angles.

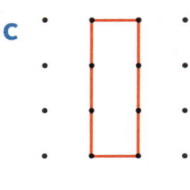
Draw diagrams to help you.

💡 Problem solving

3 A shape has 2 right angles. What is the smallest number of sides it can have?

4 James has 2 shapes. One is a regular polygon, the other is not. Together the shapes have 5 right angles and less than 10 sides. What 2 shapes could he have?

5 A shape has less than 6 sides and one or more right angles. What shape could it be?

6 A shape has at least one right angle. It is not a hexagon and it does not have 4 sides. What shape could it be?

Quadrilaterals

All polygons with 4 sides are called quadrilaterals.

Some quadrilaterals have special names because they have some **properties** that other quadrilaterals don't have.

Look at these quadrilaterals.

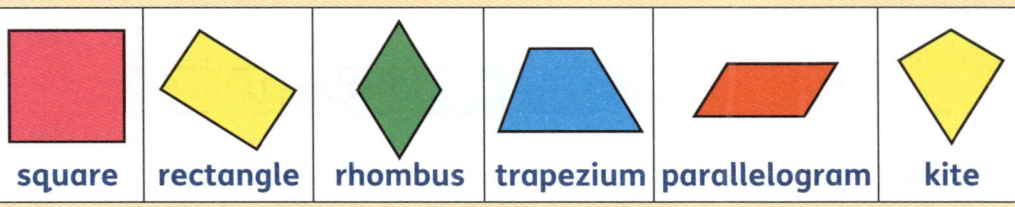

| square | rectangle | rhombus | trapezium | parallelogram | kite |

Can you work out the special properties of each one?

1 Zahra drew these quadrilaterals on dotted paper.

Write the name of each shape.

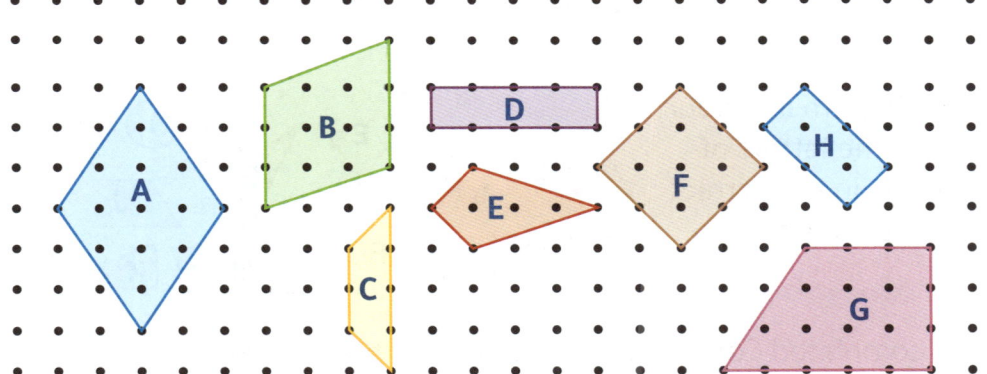

2 Read the description and then write the correct name of each shape.

a a shape with 4 sides

b a regular quadrilateral

c a quadrilateral with opposite sides the same length and 4 right angles

d a quadrilateral with 4 sides the same length but no right angles

e a quadrilateral with opposite sides the same length and no right angles

➡ *Workbook page 18*

More about quadrilaterals

A **Carroll diagram** is a table that uses columns and rows to sort information.

Each box in a Carroll diagram has two rules – the rule at the top of the column, and the rule at the beginning of the row.

When you put something into a box in the diagram, it must fit both rules.

Look at this Carroll diagram.

	Purple	Not purple
Quadrilateral		
Not quadrilateral		A

Shape A is not purple *and* it is a not a quadrilateral.

Can you work out the rules for the other three boxes in the table?

1. Copy the diagram into your book. Write the letters of these shapes in the correct places in the diagram.

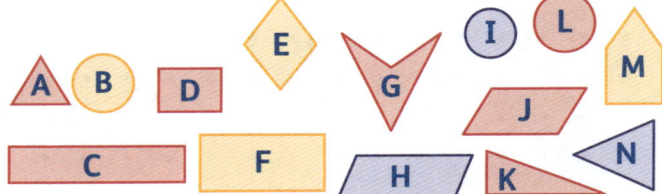

2. Use your completed Carroll diagram to answer these questions.

   a How many purple shapes are not quadrilaterals?

   b How many shapes are not purple and not quadrilaterals?

3. Work in groups to *discuss* these questions.

   a Why can you say that a rectangle is a special type of parallelogram?

   b Why are all squares also rectangles?

   c Is it true that a square is also a type of:
   - parallelogram
   - rhombus?

   d Which quadrilaterals have 4 equal sides?

Investigate squares and rectangles

You will need dotted paper, coloured pens and a ruler.

1 Draw a frame around 9 dots on your dotted paper like this:

    a Use different-coloured markers to see how many different-sized squares you can make using the 9 dots. Each corner of a square must be on a dot.

    b How many different-sized rectangles (not including squares) can you make using the 9 dots?

2 Draw a frame round 16 dots on your paper like this:

    a How many different-sized squares can you make using the 16 dots?

    b How many different-sized rectangles (not including squares) can you make using the 16 dots ?

3 Use a fresh piece of dotted paper to draw some of the squares and rectangles you made on the 9-dot and 16-dot **grids**. This time, make sure the shapes do not overlap.

    a Draw diagonals on your squares and rectangles.

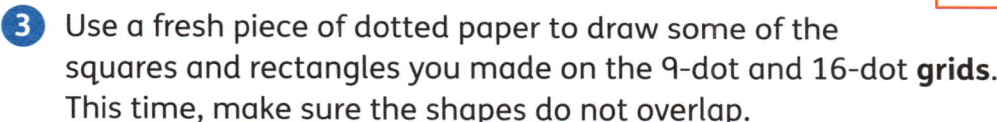

    b What do you notice about these diagonals?

A diagonal is the line that joins a corner to the opposite corner.

💡 Problem solving

Use real objects to model the shape patterns.

4 Make these stick patterns and solve the puzzles. Draw the new pattern.

    a

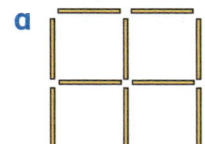

Remove 2 sticks to leave just 2 squares.

    b

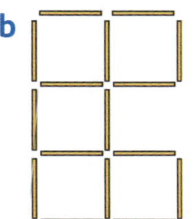

Move 3 sticks to make 4 squares.

    c

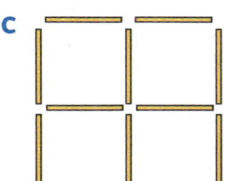

Move 2 sticks to make 7 squares.

Analogue clocks

> ### 💭 Think and share
>
> An **analogue** clock has numbers, and hands that move to show the time.
>
> - What is the difference between the hands and how they move?
>
> - How do you use both hands on an analogue clock to tell the time?

1 Which number on the clock face does the long hand (the minute hand) point to when it is showing each of these times?

   a half past the hour   **b** quarter to the hour

   c quarter past the hour   **d** 25 minutes past the hour

   e 10 minutes past the hour   **f** five minutes to the hour

2 Write each time in words.

a

b

c

d
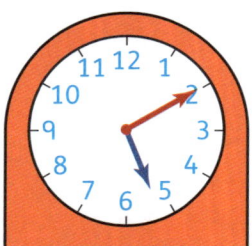

e

f

➡ *Workbook page 19*

Digital clocks

Today, most clocks are **digital**.

08:15 is 15 minutes past 8 o'clock or quarter past 8.
08:35 is 35 minutes past 8 o'clock or 25 minutes to 9 o'clock.

Some digital clocks also show the letters **a.m.** or **p.m.** after the time.

a.m. means before noon. These are times from midnight to midday.
06:00 a.m. is 6 o'clock in the morning.

p.m. means after noon. These are times from midday to midnight.
06:00 p.m. is 6 o'clock at night.

1 Write the time shown on each digital clock in words.

a   **05:20** a.m.

b   **03:30** a.m.

c   **07:00** a.m.

d   **12:05** a.m.

e   **12:45** p.m.

f   **07:40** a.m.

g   **11:35** p.m.

h   **04:40** a.m.

i   **09:45** p.m.

💡 **Problem solving**

2 Find matching pairs of times. Write each time as it would appear on a digital clock.

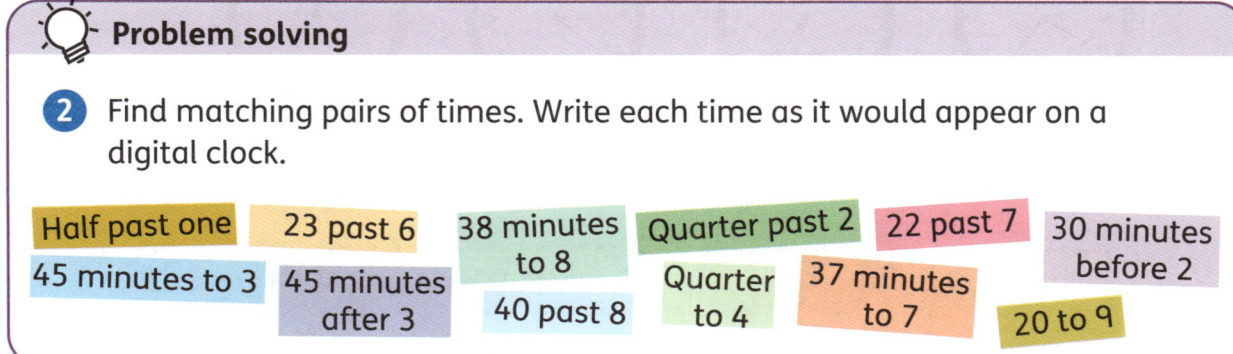

| Half past one | 23 past 6 | 38 minutes to 8 | Quarter past 2 | 22 past 7 | 30 minutes before 2 |

45 minutes to 3   45 minutes after 3   40 past 8   Quarter to 4   37 minutes to 7   20 to 9

➡ *Workbook page 19*

24-hour notation

Instead of counting hours from 1 to 12 in the morning and 1 to 12 in the afternoon, we can count from 1 to 24.

This is called **24-hour notation**.

Before noon (a.m.)

12 midnight until 11.59 a.m.

00:00 until 11:59

After noon (p.m.)

12 noon until 11.59 p.m.

12:00 until 23:59

1 Write each time in words. The first one has been done for you.

a 13:30 *Half past one in the afternoon*

b 23:45 **c** 00:00 **d** 04:15 **e** 17:25

f 19:55 **g** 01:10 **h** 20:00 **i** 22:30

2 Write each time using 24-hour notation. The first one has been done for you.

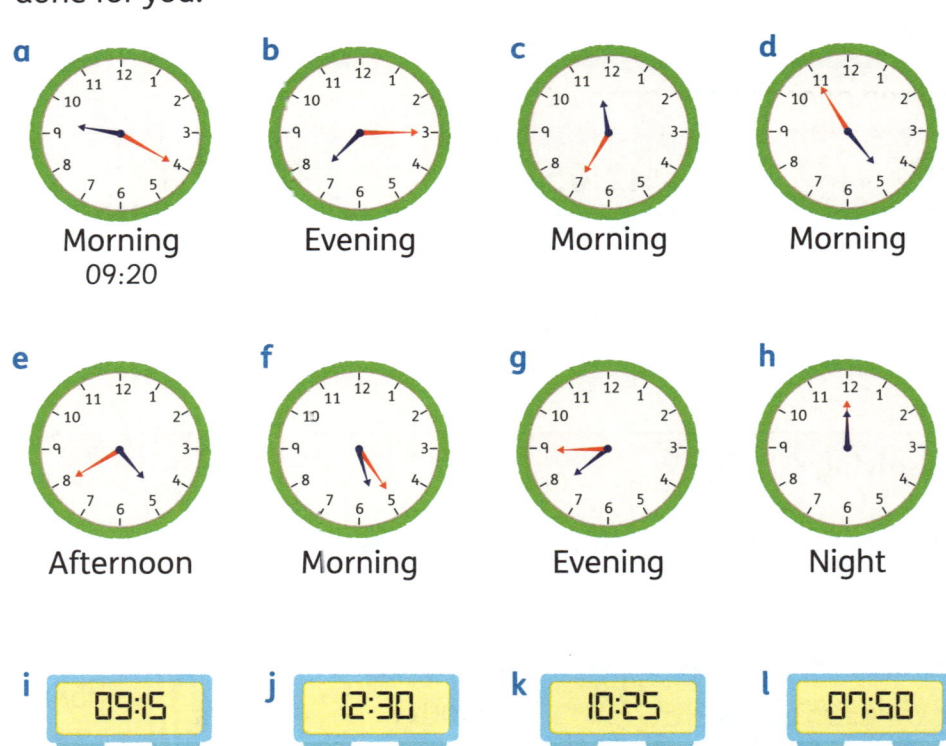

a Morning 09:20 **b** Evening **c** Morning **d** Morning

e Afternoon **f** Morning **g** Evening **h** Night

i 09:15 Evening **j** 12:30 Midday **k** 10:25 Evening **l** 07:50 Evening

➡ *Workbook page 20*

Timetables

This bus timetable shows eight different bus stops and the time the bus arrives at each stop.

Bus Timetable – Bus A232 Route 6	
Central bus station	7:15
Klimt Street	7:40
Rembrandt Square	7:55
Da Vinci Boulevard	8:03
Picasso Place	8:20
Mondrian Avenue	8:42
Escher Street	8:55
Rubens Road	9:10

1 At what time does the bus arrive at:

   a the Central bus station **b** Picasso Place   **c** Rubens Road?

2 At which stop does the bus arrive at:

   a 7:40   **b** 8:42   **c** five to nine?

3 If the bus is 10 minutes late, what time does it arrive at:

   a Klimt Street   **b** Escher Street   **c** Rembrandt Square?

4 If the bus is 10 minutes early, where will it be at these times?

  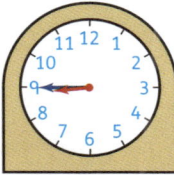

5 How long does it take the bus to travel:

   a from the Central bus station to Klimt Street

   b from Da Vinci Boulevard to Mondrian Avenue

   c from the Central bus station to Rubens Road?

➡️ *Workbook page 21*

Reading timetables

Timetables often use the 24-hour notation.

Here is a timetable for four buses.

Arriving at:	Bus A12	Bus C15	Bus C19	Bus D23
Drummond Street	09:46	11:46	13:46	16:46
Fifth Avenue	10:07	12:07	14:07	17:07
King Street	10:39	12:39	14:39	17:39
Ajman Road	11:16	13:16	15:16	18:16
Hill Street	11:49	13:49	15:49	18:49
Fort Avenue	12:31	14:31	16:31	19:31
Emir Lane	13:04	15:04	17:04	20:04

1 Answer these questions about the timetable.

   a What time does bus A12 arrive at Drummond Street?

   b Which buses arrive at Drummond Street later than 1 p.m.?

   c When does bus C19 arrive at Fort Avenue?

   d Cara catches a bus at Hill Street at 11 minutes to 4.
   Which bus is this?

2 Yolandi catches Bus C15 from King Street to Emir Lane.

   a What time does the bus arrive at King Street?

   b How long does it take to travel from King Street to
   Ajman Road?

   c Does bus CI5 arrive at Emir Lane before or after 3 p.m.?

   d One day Yolandi decides to take Bus C19 instead of C15 from
   King Street to Emir Lane. How much later does Yolandi arrive
   at Emir Lane?

   e Yolandi wants to be home by 3.20 p.m. to watch a TV
   programme. The C15 bus arrives at Emir Lane 3 minutes late.
   After she gets off the bus at Emir Lane it takes her 8 minutes
   to walk home. Will she make it on time?

Converting units of time

How many minutes are there in 3 hours?

Multiply 3 hours by the number of minutes in 1 hour.

1 hour = 60 minutes   3 hours = 3 × 60 minutes = 180 minutes

A timer shows 180 seconds. How many minutes is this?

Divide 180 seconds by the number of seconds in 1 minute.

60 seconds = 1 minute

In minutes, 180 seconds = 180 ÷ 60

= 3 minutes

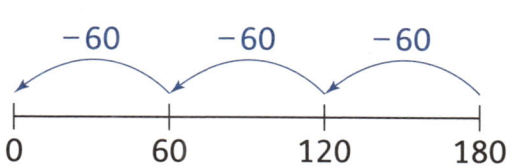

The diagram shows how units of time are related to each other.

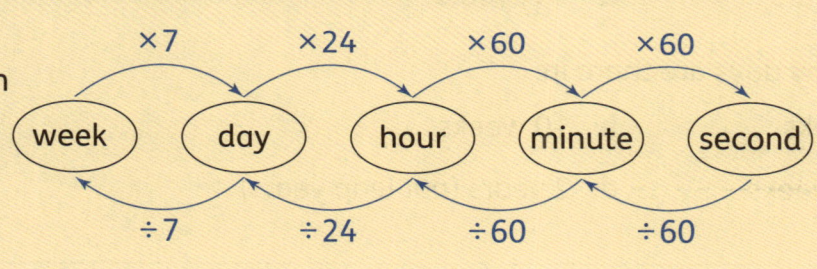

Multiply to convert from a larger unit to a smaller unit.

Divide to convert from a smaller unit to a larger unit.

1. Convert these times. Show your working out.

    a   3 days to hours
    b   4 hours to minutes
    c   5 minutes to seconds
    d   10 days to hours
    e $3\frac{1}{2}$ hours to minutes
    f $4\frac{1}{2}$ minutes to seconds
    g   48 hours to days
    h   90 minutes to hours

Problem solving

Think about which operations to use.

2. A machine takes 3 hours to make 30 T-shirts.
   How many minutes does it take to make 1 T-shirt?

3. Solly finished a race in 1 minute and 50 seconds. Musa finished in 94 seconds. How many seconds faster was Musa?

4. A farmer spends 1 week, 2 days and 7 hours planting crops. How many hours is this?

More time conversions

A year is a unit of time. You can also measure years in days, weeks or months.

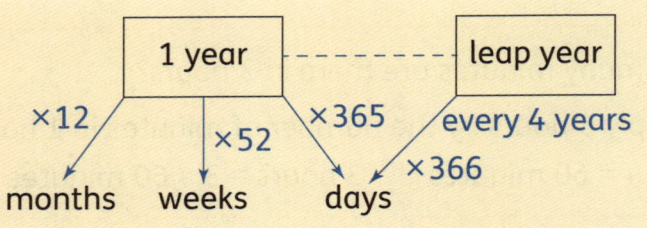

1 Describe each of these using different units of time.

   a 12 months   b 14 days

   c 6 months   d 366 days

2 How many months are there in:

   a 2 years   b $\frac{1}{2}$ a year

   c $3\frac{1}{4}$ years   d $4\frac{7}{12}$ years

3 How many days are there in:

   a 2 weeks   b 10 weeks

   c 100 weeks   d 2 years (*not* leap years)

 Problem solving

4 Sanita works out that she goes to school for 32 weeks in a year. How many school days is this?

5 Farid sleeps 8 hours per night. How many hours does he sleep in:

   a a week   b a year?

6 Eva is a swimmer. She trains for 50 minutes after school every day.

   a How many hours does she train per week?

   b How many hours will Eva train during the month of February? (There are 4 weekends and no school holidays in February.)

   c There are 35 school weeks in a year. How long does Eva spend training in one school year? Give your answer in two different units of time.

➡ *Workbook page 22*

UNIT 5 Decimals

Decimals and fractions

💭 Think and share

We use **fractions** and **decimals** in different ways in our daily lives.

Where might you see these examples?

Fractions	Decimals

Where else have you seen fractions and decimals around you? Tell your group.

We can use decimals to write money amounts.

£1.25 means 1 whole pound and $\frac{25}{100}$ of a pound (you say 'twenty-five **hundredths** of a pound'). This is 1 pound and 25 pence.

$2.50 means 2 whole dollars and $\frac{50}{100}$ of a dollar. This is 2 dollars and 50 cents.

• Which of these is the correct way of writing one pound and twenty-seven pence? Why are the other ways wrong?

   £127   £12.70   £1.27   £0.127   £127.00

1 Write these money amounts as decimals.

   a Two pounds and thirty pence

   b Six pounds and ninety-three pence

   c Twelve pounds and five pence

   d Four pounds

2 How many pence are in each of these amounts?

   a £4.87   b £5.95   c £12.30   d £2.05   e £6.00

3 Write these amounts in order from most money to least money.

   99p   £9.00   £0.90   £1.99

Decimal place value

Each big square represents 1 whole. A fraction of each square is shaded.

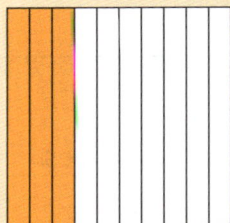

$\frac{3}{10}$ of this square is shaded. $\frac{25}{100}$ of this square is shaded.

The shaded part of each square is less than 1, so we write it to the right of the ones place on the place-value table.

We write tenths and hundredths in the place-value table like this:

Hundreds (100)	Tens (10)	Ones (1)	.	tenths $\left(\frac{1}{10}\right)$	hundredths $\left(\frac{1}{100}\right)$
		0	.	3	
		0	.	2	5

$\frac{3}{10}$ can be written as 0.3 and $\frac{25}{100}$ can be written as 0.25

We write a 0 as a placeholder in the ones place because these decimals have no whole-number parts.

1 Write the unshaded part of each square in the example above as a fraction and as a decimal.

2 Each shape represents one whole. Write a fraction and a decimal for the shaded part of each shape.

a 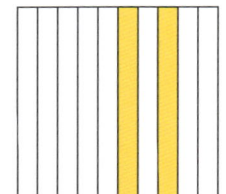 b 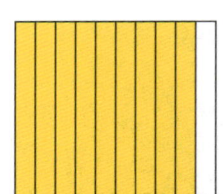 c 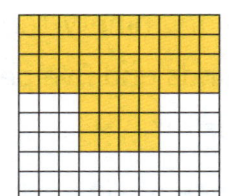 d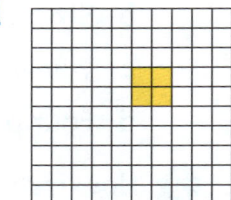

3 What is the value of the underlined red digit in each number?

a 12.3<u>4</u>   b <u>1</u>8.98   c 13.0<u>8</u>   d 9.9<u>9</u>   e <u>3</u>.45

Workbook page 23

Modelling decimals

Here are some ways of showing decimals.

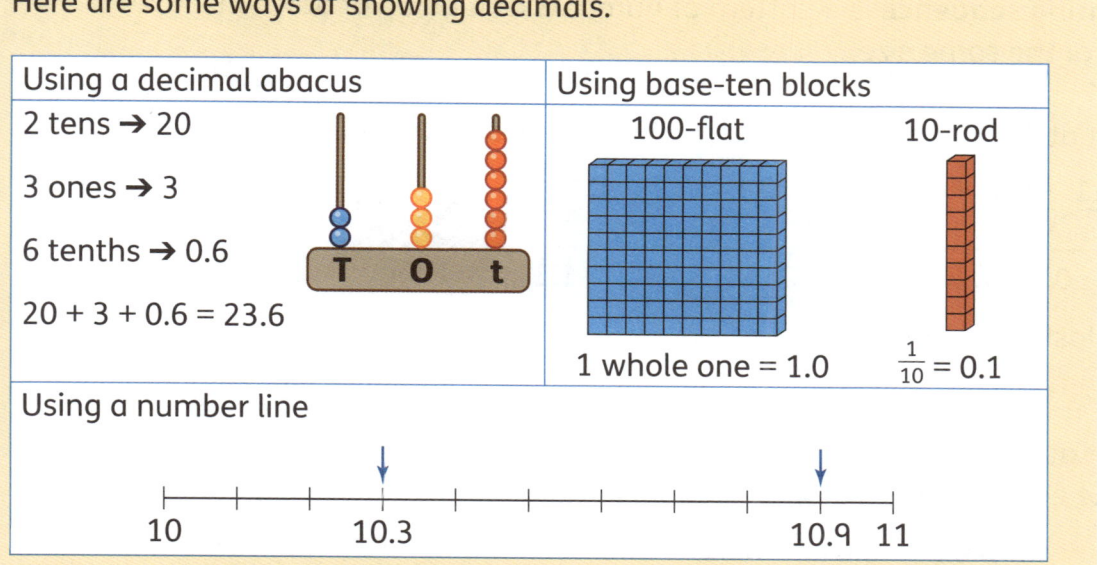

Using a decimal abacus	Using base-ten blocks

2 tens → 20

3 ones → 3

6 tenths → 0.6

20 + 3 + 0.6 = 23.6

100-flat   10-rod

1 whole one = 1.0 $\frac{1}{10}$ = 0.1

Using a number line

10   10.3   10.9 11

1 Write the decimal shown by each group of base-ten blocks.

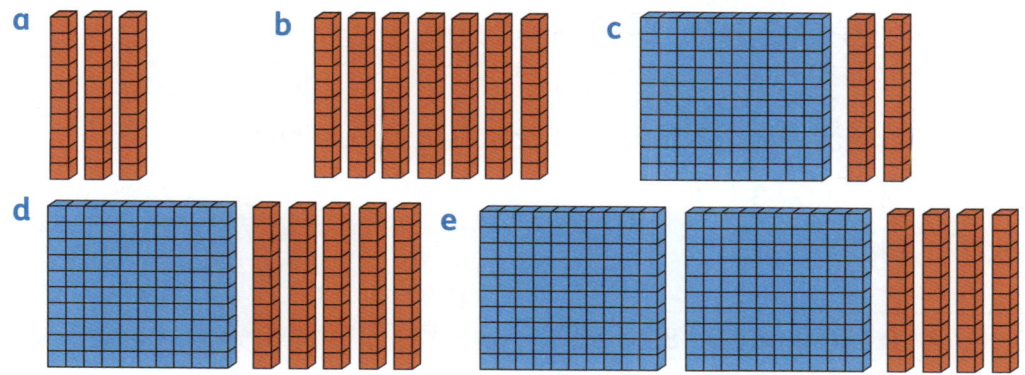

a   b   c

d   e

2 Write the decimal shown on each abacus.

a 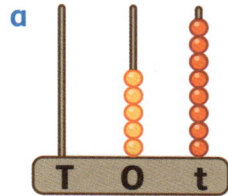   b 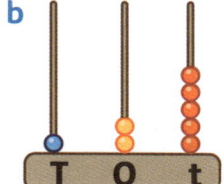   c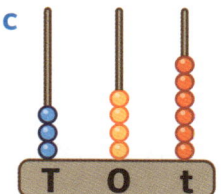

3 Write the number shown by each arrow on this number line.

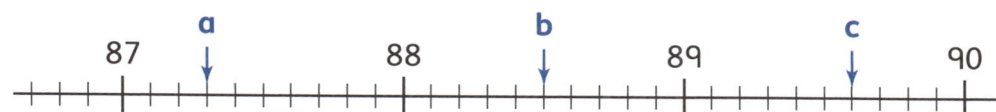

a   b   c

87   88   89   90

Count in tenths and hundredths

A **counting sequence** is a pattern of numbers that go up or down in jumps of the same size.

Start at 0.6 and count on 7 tenths. Write the last number you count.

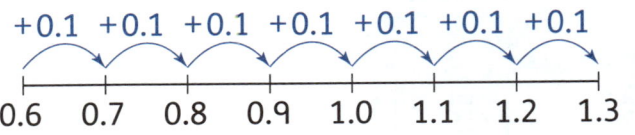

+0.1 +0.1 +0.1 +0.1 +0.1 +0.1 +0.1

0.6   0.7   0.8   0.9   1.0   1.1   1.2   1.3

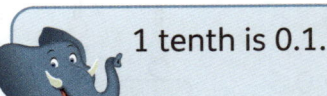

1 tenth is 0.1.

The last number is 1.3.

Start at 5.25 and count back 3 hundredths.
Write each number you count.

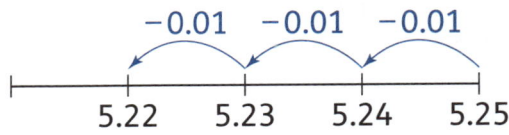

−0.01   −0.01   −0.01

5.22   5.23   5.24   5.25

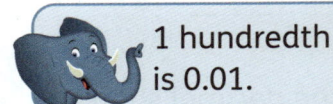

1 hundredth is 0.01.

The numbers are 5.25, 5.24, 5.23, 5.22.

What numbers are missing from this counting sequence?

0.40, 0.42, ☐, ☐, 0.48, ☐

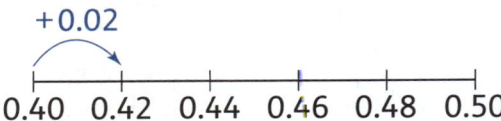

+0.02

0.40   0.42   0.44   0.46   0.48   0.50

The missing numbers are 0.44, 0.46 and 0.50.

1 Write the next four numbers in each of these counting sequences.

   a 3.1, 3.2, 3.3, ...   **b** 8.6, 8.7, 8.8, ...

   c 12.8, 12.6, 12.4, ...   **d** 4.61, 4.60, 4.59, ...

   e 2.83, 2.85, 2.87, ...   **f** 4.35, 4.3, 4.25, ...

2 What number do you reach when you:

   a count on 9 hundredths from 12.95

   b count back 7 hundredths from 0.83?

➡ *Workbook page 24*

More tenths and hundredths

1 Write down the missing numbers in each of these counting sequences. Check your answers using a calculator.

a $2 \xrightarrow{+0.5} \square \xrightarrow{+0.5} \square \xrightarrow{+0.5} \square \xrightarrow{+0.5} \square \xrightarrow{+0.5} \square \xrightarrow{+0.5} \square$

b $7 \xrightarrow{-0.2} \square \xrightarrow{-0.2} \square \xrightarrow{-0.2} \square \xrightarrow{-0.2} \square \xrightarrow{-0.2} \square \xrightarrow{-0.2} \square$

c $0 \xrightarrow{+0.25} \square \xrightarrow{+0.25} \square \xrightarrow{+0.25} \square \xrightarrow{+0.25} \square \xrightarrow{+0.25} \square \xrightarrow{+0.25} \square$

d $90 \xrightarrow{-0.05} \square \xrightarrow{-0.05} \square \xrightarrow{-0.05} \square \xrightarrow{-0.05} \square \xrightarrow{-0.05} \square \xrightarrow{-0.05} \square$

2 Write the number shown by each arrow on the number lines.

```
         a    b    c              d    e    f
         ↓    ↓    ↓              ↓    ↓    ↓
|———+————+————+————+————|    |—+—+—+—+—+—+—+—+—+—|
4.2  4.4                5.2   2.98                3.08
```

Problem solving

> Draw number lines if you need them.

3 At the end of summer, a shop owner decides to reduce the price of summer items by 0.4 pounds. What will the new price of each of these items be?

a  £4.80

b £2.20

c  £1.05

d £4.00

e  £9.25

---

4 Usain Bolt ran 100 m in 9.58 seconds to set a world record.

a The second-place runner was $\frac{14}{100}$ of a second slower. What was that time?

b Another athlete, Shelly-Ann Fraser-Pryce, won the 100 m world championships in 2019 with a time of 10.17 seconds. How many seconds faster would she need to run to match 9.58 seconds?

Compare and order decimals

We can use place value to compare the size of decimals.

Compare these decimals. Write <, = or > in place of 'and'.

19.27 and 19.72

1 9.2 7
1 9.7 2
↑ 7 > 2
So 19.72 > 19.27

3.1 and 3.01

3.1 0 ← 0 as placeholder
3.0 1
↑ 1 > 0
So 3.1 > 3.01

1   Compare these decimals. Write <, = or > in place of 'and'.

   a   0.26 and 0.52   b   1.3 and 1.30   c   0.07 and 0.7

2   Write each set of decimals in order from smallest to greatest.

   a   8.81   8.18   8.8   8.17
   b   9.96   9.92   10.00   10.80   9.99   9.58

Make an
organised list.

 Problem solving

3   This diagram shows how far Sanjay, Amira, Musa, Julia and Kim travel
    to school each day, but it does not say who travels each distance.

    Read the clues and match each child to the distance they travel to school.

- Sanjay travels less than Amira but more than Musa.
- Julia travels less than 6 km.
- Amira travels more than 6 km but less than Kim.
- Musa doesn't travel as far as Julia.

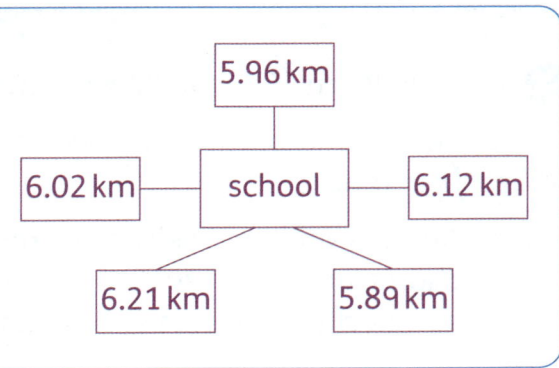

➡ *Workbook page 25 and page 26*

Round decimals to the nearest whole number

To round a decimal to the nearest whole number, look at the digit in the tenths place.

- If this digit is less than 5, the whole number does not change.
- If this digit is 5 or more, you round up to the next whole number.

| 1 | 1.4 rounds to 1 | 1.5 | 1.5 rounds to 2 | 2 |

If tenths place is less than 5, number stays the same

If tenths place is 5 or more, round up

1 Round each decimal to the nearest whole number.

   a 0.9   **b** 1.4   **c** 9.6   **d** 5.4   **e** 12.8   **f** 19.9

2 Round each decimal to the nearest whole number.

   a 164.2   **b** 234.7   **c** 765.4

   d 543.4   **e** 515.8   **f** 599.9

3 Nathi rounded six masses to the nearest whole kilogram and got these results:

   25 kg   26 kg   24 kg   20 kg   29 kg   30 kg

   a Write down Nathi's rounded measurements.

   b Match the measurements in the box to the kilogram amounts they round to.

28.5 kg	25.2 kg	24.5 kg	29.8 kg	30.3 kg
25.6 kg	25.3 kg	24.8 kg	19.9 kg	20.3 kg
25.9 kg	26.4 kg	29.4 kg	29.5 kg	25.8 kg	24.4 kg

4 A delivery van drove 39.5 km on Monday, 123.2 km on Tuesday and 59.9 km on Wednesday.

   a Round each distance to the nearest **kilometre**.

   b Approximately how many kilometres did the van travel altogether over the three days?

Solve problems involving decimals

There are four main events in a gymnastics competition.
The table shows five gymnasts' scores in each event.

	Vault	Bars	Floor	Beam
Francesca	8.50	7.65	8.00	7.15
Barbara	7.85	8.35	8.25	7.40
Mino	8.25	7.80	8.50	8.20
Simone	8.10	8.55	9.10	8.80
Olga	7.95	8.25	8.75	8.50

1 Find the following information in the table:

   a the highest score for each event

   b each gymnast's highest score.

2 Order each gymnast's scores for the four events from highest to lowest.

3 Who do you think won the gold, silver and bronze medals in this competition? Explain how you decided this.

4 Here are the scores of four of the gymnasts in another competition:

Francesca	Barbara	Mino	Simone	Olga
9.8	9.76	9.79	9.82	

   a What is the lowest score (to the nearest hundredth) that Olga can get in order to win?

   b If Olga wins, who comes second and third?

Measures and money

Metric units of length

> ### Think and share
>
> The diagram shows two lines that are 1 **centimetre** apart.
>
> 1 centimetre can also be written as 1 cm.
> cm is short for centimetre. 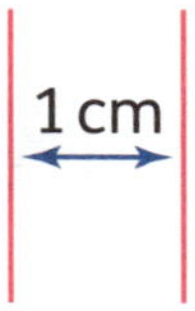
>
> 1 cm = 10 millimetres (mm)
> 100 cm = 1 metre (m)
> 1000 m = 1 kilometre (km)
>
> 1 cm
>
> Millimetres, centimetres, metres and kilometres are metric units of **length**.
>
> - Which of these units would you not find on your ruler? Why?
>
> | handspan | millimetre | kilometre | **litre** | centimetre | metre |
> |----------|------------|-----------|-----------|------------|-------|
>
> - Matteo says length and height are different measurements.
>   How could you convince him that they are the same?

1. Work in pairs to decide what the missing units of length are in each of these sentences.

    a Sarah is 11 years old and 142 _____ tall.

    b Marcel drove 68 ___ to the next town.

    c The point on a sharp pencil is 5 _____ long.

    d Anita needed a 10 _____ strip of card to finish her model.

    e My fingernail is 9 _____ wide.

    f Abdul's mobile phone is 8 _____ thick.

    g Nita ran a 100 _____ race.

    h Khalid and Sara walk about 225 _____ to school.

    i In 2019, Brigid Kosgei of Kenya finished the 42.2 __
    Chicago Marathon and set a new world record time.

2. Use a ruler and pencil to draw lines that are:

    a 4 cm in length   b 70 mm long   c 2.5 cm tall

➡ Workbook page 27

Read and write lengths

Julie measured her height. She was 1 m and 19 cm tall. Julie's height can be written in three ways:

1 m 19 cm or 119 cm or 1.19 m.

Look at this ruler. The centimetres are numbered and shown with a longer line.

The millimetres are shown, but they are not numbered. You have to count them to work out how long the green line is.

The green line is 2 cm and 9 mm long. This is the same as 2.9 cm or 29 mm.

1 Measure your own height. Write it in three ways.

2 Write each of these measurements in three different ways.

a

The car is 250 cm long.

b

The table is 1 metre and 47 cm long.

3 What is the length of each of these green lines? Write each length in three different ways.

➡ *Workbook page 28*

Litres and millilitres

The **capacity** of a container tells us how much it can hold. The standard metric unit of capacity is the litre.

The drawings show the capacity of some common measuring instruments.

1 litre (1 ℓ) = 1000 millilitres (1000 ml)

Measuring spoons and cups have exact capacities.

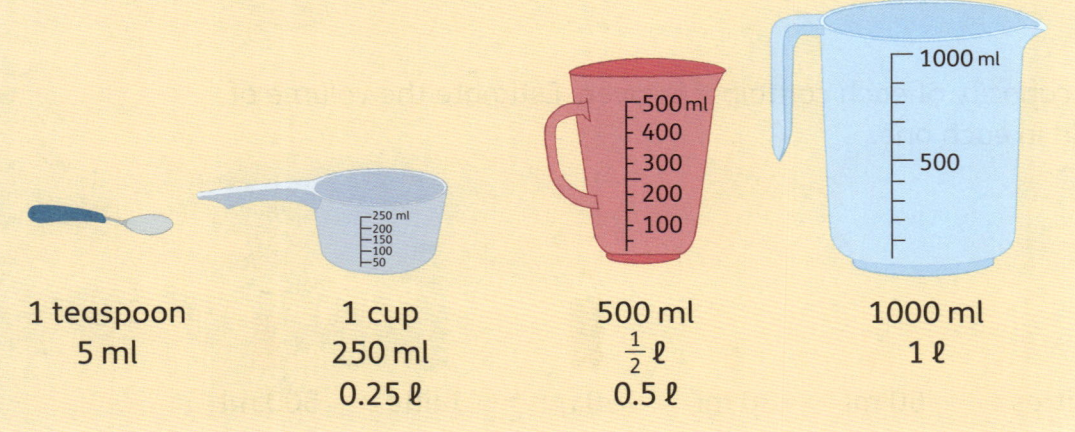

1 teaspoon	1 cup	500 ml	1000 ml
5 ml	250 ml	$\frac{1}{2}$ ℓ	1 ℓ
	0.25 ℓ	0.5 ℓ	

1 Collect some containers used for yoghurt, juice, milk and other liquids.

   a Arrange your containers in order from the one with the smallest capacity to the one with the greatest capacity.
   Try to estimate the capacity of each container.
   b Use measuring jugs and spoons to measure the exact capacity of each container.

2 Use the capacities shown above to work out the answers to these questions.

   a How many teaspoons are there in 1 cup?
   b How many cups are there in 1 litre?
   c How many cups are there in 2.5 litres?

3 How many litres do you need to fill:

   a 8 cups   **b** 9 cups   **c** 11 cups   **d** 22 cups?

Estimate and calculate amounts of liquid

You learnt that the capacity of a container is how much it can hold.

Containers are not always filled to capacity.

This bottle has a capacity of 1 litre but it only has $\frac{1}{4}$ of a litre of liquid inside it.

Volume is how much space something takes up. The volume of liquid in the bottle is $\frac{1}{4}$ litre. This is the same as 250 millilitres.

1 The *capacity* of each container is given. Estimate the *volume* of liquid in each one.

| 20 litres | 50 ml | 30 ml | 750 ml | 1 litre | 500 ml |

2 A scientist made some liquid mixtures.

   a Calculate the total volume of liquid in each of these mixtures.

Mixture A	Mixture B
250 ml lemon juice	350 ml milk
175 ml water	50 ml lemon juice
50 ml oil	150 ml oil
	200 ml ink
Mixture C	**Mixture D**
150 ml lemon juice	550 ml milk
250 ml oil	250 ml lemon juice
220 ml water	350 ml oil
125 ml ink	425 ml water
	375 ml ink

   b The scientist started with 1 litre of each of these liquids: milk, lemon juice, oil, ink and water.

Calculate the volume of each type of liquid left after the mixtures are made. Give your answers in millilitres.

➡ *Workbook page 29*

Calculate with measures

Problem solving

Read these problems carefully and then solve them. Show your working clearly.

> Draw diagrams to represent the problems. An example is given for the first problem.

1 The instructions on a 250-ml bottle of medicine say 'Take 10 ml twice a day'. How long will the bottle last?

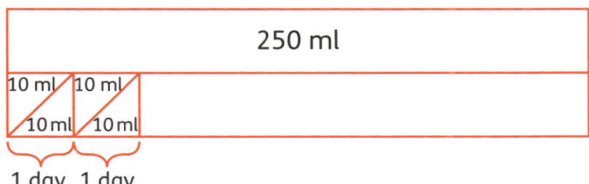

2 A plane uses 75 litres of fuel for every hour that it flies. How much fuel will it use on a flight lasting 3 hours and 20 minutes?

3  Solly puts a 12-litre bucket under a leaky tap to collect the water. 1500 millilitres of water drip from the tap each hour. How long will it take to fill the bucket?

4 A small car uses 1.25 litres of petrol to travel 20 km. The petrol tank holds 35 litres. How far can the car travel on a full tank of petrol?

5 A health store sells fresh juice for £1.20 per litre. Customers bring their own containers and fill them. How much should the store charge if the customer fills:

a a 2-litre bottle

b two bottles, each with a capacity of 1.5 litres

c a 500-ml bottle

d a 400-ml plastic cup?

6 How could you measure exactly 4 litres of water using these 3 buckets?

7 ℓ   5 ℓ   3 ℓ

> Use a **bar model** or make an organised list to help you keep track.

Mass

You measure the mass of an object to find out how heavy or light it is.
Kilograms (kg) and **grams** (g) are metric units of mass: 1 kg = 1000 g.
These instruments measure mass.

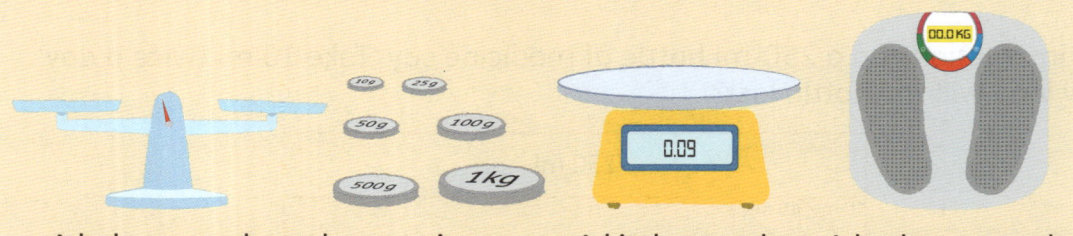

A balance scale and mass pieces   A kitchen scale   A bathroom scale

1 Convert each of these kilogram measurements to grams.

    a 2 kg   b 25 kg   c 1.5 kg   d 1.75 kg   e $8\frac{1}{2}$ kg

2 Round each of these measurements to the nearest whole kilogram.

    a 2.76 kg   b 3.29 kg   c 35.09 kg   d 142.24 kg

3 Look at these items.

A   B   C

D   E   F

    a Arrange the items in order from heaviest to lightest.
      Write the letters one below the other in your book.

    b Next to each letter, write whether you would measure the
      mass of the item in kilograms or grams.

    c Estimate the mass and write it next to the letter.

    d Discuss how you could check how accurately you estimated.

➡ *Workbook page 30*

Read measuring scales

Look at this scale.

The kilograms are numbered.

There is a mark halfway between each pair of numbers. These marks show $\frac{1}{2}$ kg or 500 g.

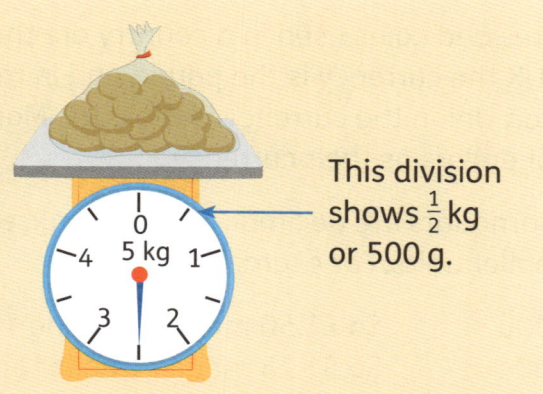

This division shows $\frac{1}{2}$ kg or 500 g.

These scales show these potatoes have a mass of 2 kg and 500 g.

You can write this in three ways:
2 kg 500 g or 2500 g or 2.5 kg.

1 What mass is shown on each of these scales?

a
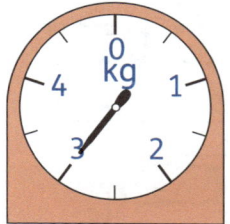

b
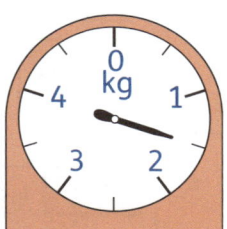

c
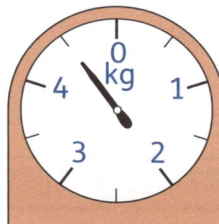

d
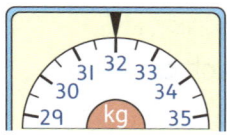

e

f
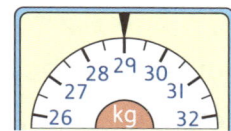

2 Work with a partner.
Look at these scales.

a What does each small mark represent?

b Write the measurement shown by each letter (A to F).

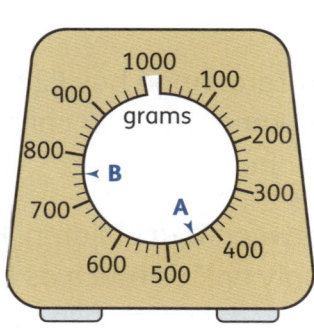

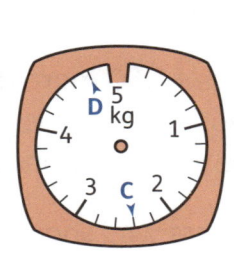

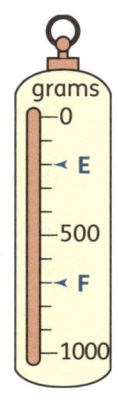

➡ *Workbook page 31*

Money

The notes and coins used in a country are the **currency** of that country. In the UK the currency is the pound and in most countries in the European Union the currency is the euro. Many countries use different types of dollars as their currency.

Money amounts are decimal. The main currency unit is divided into 100 smaller units. For example:

£1 = 100p	So £1.50 means 1 pound + 50 pence or 150 pence
$1 = 100c	So $1.25 means 1 dollar and 25 cents or 125 cents
€1 = 100c	So €1.90 means 1 euro and 90 cents or 190 cents

1 Write the total amount of money shown in each box as a decimal.

a 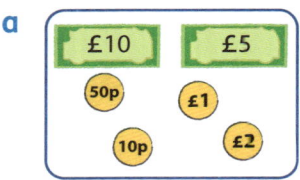 b 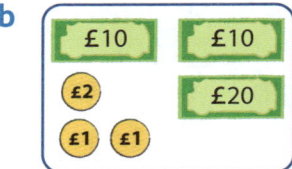 c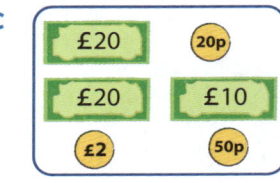

d 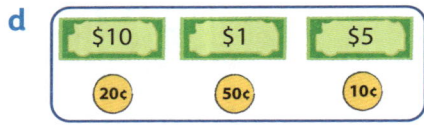 e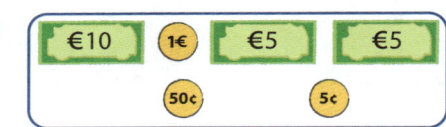

2 What is the money value of the underlined red digit in each amount?

a £5̲3.35	b £26.5̲8	c £27.05̲
d €1̲4.50	e €11.6̲5	f €42.1̲0
g $124̲.67	h $4̲50.90	i $24̲5.95

Draw a bar model.

 Problem solving

3 Rabia has £1.00 more than Zara. Anna has £2.00 more than Zara. Maria has £3.00 more than Anna. The four children have £20.00 in total. How much money do they each have?

4 Fifty dollars was shared between four children. Kevin got double the amount that Dali got but only half as much as Nick. Sanjay got $5 more than Kevin. How much money did each child get?

➡ Workbook page 32

Calculate money amounts

We often need to calculate with money amounts in our daily lives.

- We calculate how much it will cost to buy things.

- We calculate how much change we will get when we pay.

> Change the dollar amounts to cents so you can work with whole numbers instead of decimals.

A ticket for a ferry costs $4.25. How much change will you get if you pay with a $5 note?

$5 = 500c
$4.25 = 425c

500c – 425c = 75c
You will get 75 cents change.

Carrots cost 60p a bunch. How much would four bunches cost?

$$4 \times 60 = 4 \times 6 \times 10$$
$$= 24 \times 10$$
$$= 240p$$

Four bunches would cost £2.40.

1 Zara has two £1 coins, a £2 coin, and seven 20p coins. How much money does she have altogether?

2 Naresh has $15.75. He wants to buy a game that costs $19.99. How much more money will he need?

3 Avi bought snacks costing $4.85 and got $4.15 change from a $10 note. Is this correct or not?

4 After buying a book for £2.80 and a pen for 95p, Silvio had £4.23 left. How much money did he start with?

5 Jessica saves $4.25 every week for 12 weeks. How much will she have at the end of the twelfth week?

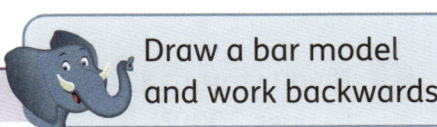

> Draw a bar model and work backwards.

Problem solving

6 Salman gives half of the money in his wallet to his brother and spends one-third of the **remainder**. He has 90 pence left. How much did he start with?

Count and calculate

Count on and back

> ### 💭 Think and share
>
> • Which digits change in each of these counting sequences? Why?
>
> • What will happen to the other digits as you keep counting? Why?
>
> | Count on in ones: | 1234 | 1235 | 1236 |
> |---|---|---|---|
> | Count on in tens: | 1234 | 1244 | 1254 |
>
> | Count back in ones: | 3688 | 3687 | 3686 |
> |---|---|---|---|
> | Count back in hundreds: | 3688 | 3588 | 3488 |
>
> • What is the counting rule for each of these counting sequences?
>
>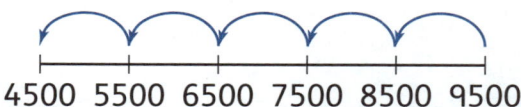
>
> 3700 4700 5700 6700 7700 8700   4500 5500 6500 7500 8500 9500

1 Write the next five numbers in each counting sequence. You need to work out what the rule is first.

   a 250   260   270   **b** 1120   1110   1100

   c 12 343   11 343   10 343   **d** 23 456   24 456   25 456

2 There is a mistake in each **sequence**. Find it and correct it.

   a 5456   5356   5256   5156   5956

   b 12 385   12 395   12 400   12 415   12 425

   c 23 445   24 445   24 545   26 445   27 445

   d 9999   9989   9979   9978   9959

3 If you start at 1245, how many hundreds do you need to count on to pass 2000?

4 Micah says if you count in thousands, all the numbers you count will have three zeroes at the end. Do you agree or not? Explain why.

➡ *Workbook page 33*

Count in multiples of 6, 7 and 9

A multiple is the answer you get when you multiply one whole number by another.

$3 \times 6 = 18$ 18 is a multiple of 3 and a multiple of 6

$7 \times 10 = 70$ 70 is a multiple of 7 and a multiple of 10

You can count in steps to find a sequence of multiples of any number. These are multiples of 9 that you already know:

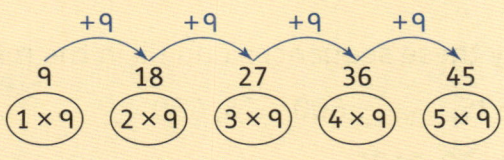

$9 \times 15 = 135$. What are the next three multiples of 9?

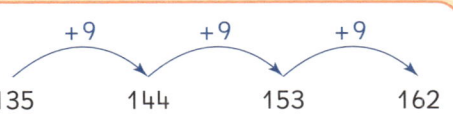

The next three multiples of 9 are 144, 153 and 162.

1   Write the next five numbers in each counting sequence.

    a   24   30   36   42   48   **b**   99   90   81   72   63

    c   35   42   49   56   63   **d**   132   121   110   99   88

2   Count on to find:

    a   the sixth multiple of 6 after 84   **b**   the tenth multiple of 7 after 70

    c   the fifth multiple of 9 after 207

Problem solving

3   What is the lowest number that is a multiple of both:

    a   6 and 9   **b**   6 and 7   **c**   7 and 9

4   Li is counting back in 7s from 161. Will she count 120?
Explain your answer.

5   Zef counted on in 6s from 100 to get this sequence.
    106   112   118   124   130
    He says these are multiples of 6. Explain why he is not correct.

➡ Workbook page 34

Count in 25s

When you count in 25s, the tens and ones repeat in a pattern.

25   50   75   100   125   150   175   200   225   250   275   300

What is the pattern?

1 Copy these sequences and fill in the missing numbers.

   a 525, __, __, 600, 625, 650

   b 1200, 1175, __, __, __

   c 1450, 1425, __, __, __

   d 23 475, __, __, 23 550, __

2 Work out the missing values. Write the letters and the values in your book.

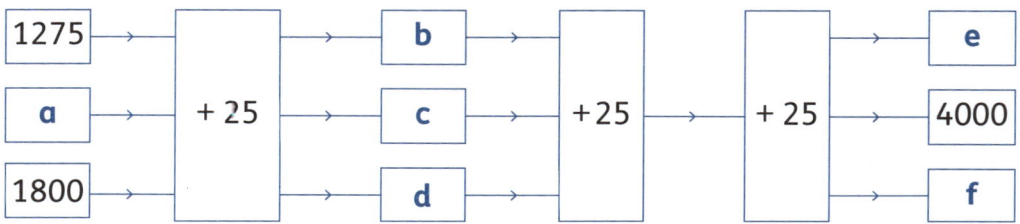

3 Zarea and Jay are counting in 25s. They both start at 175. Zarea counts back three numbers and Jay counts forwards three numbers. Which numbers will each child count?

4 Jared counted in 25s. He stopped counting at 2325. Write down the five numbers he counted before that.

5 Find the mistake in this sequence and explain what went wrong.

      1950   1975   2000   2250

Problem solving

6 How can you tell that 12 345 is not a multiple of 25, without counting?

7 Fabric trim comes in these packs:
A clothing factory needs exactly 1575 m of trim for a large order and wants to buy the fewest possible packs. What combination of packs should they buy?

PACK A	PACK B	PACK C	PACK D
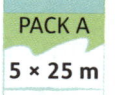 5 × 25 m	 5 × 50 m	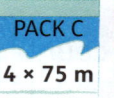 4 × 75 m	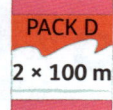 2 × 100 m

➡ *Workbook page 35*

Count in multiples to add or subtract

You can count on or back to add and subtract multiples of 10, 100 and 1000.

What is 2529 + 30?

+ 10   + 10   + 10

30 is 3 tens, so count on in tens.

2529   2539   2549   2559

2529 + 30 = 2559

What is 9600 – 4000?

– 1000 – 1000 – 1000 – 1000

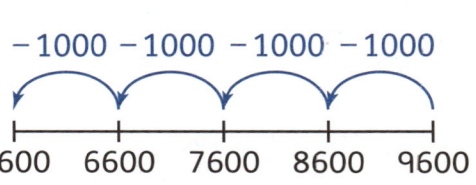

5600   6600   7600   8600   9600

1 Add.

   a 1234 + 40   b 3044 + 500   c 400 + 3560

   d 20 + 3456   e 50 + 1280   f 900 + 3200

   g 400 + 899   h 1345 + 8000   i 700 + 1200

2 Subtract.

   a 8600 – 400   b 987 – 60   c 7350 – 200

   d 9866 – 5000   e 8990 – 4000   f 6543 – 400

   g 2010 – 20   h 4345 – 400   i 5645 – 70

3 Complete the number sentences.

   a 4300 – ☐ = 4100   b 2080 – ☐ = 2030   c ☐ – 3000 = 4680

   d 580 + ☐ = 1080   e 2450 + ☐ = 2950   f 3020 + ☐ = 3100

4 Write down the number you would add to get these results:

   a 1235 → 1735   b 3448 → 3498   c 3250 → 8250

   d 4567 → 4867   e 9200 → 10 000   f 4860 → 5160

Make 100

You already know pairs of numbers that add up to 10.
You can use these to find pairs of multiples of ten that add up to 100.

$1 + 9 = 10$ → $10 + 90 = 100$ $3 + 7 = 10$ → $30 + 70 = 100$

You can also count on or count back to find any pair of numbers that add up to 100.

$36 + \boxed{} = 100$

You can think like this:

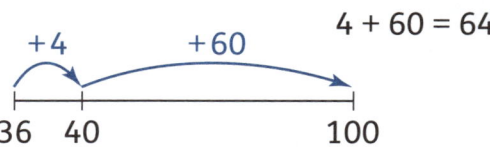

$4 + 60 = 64$

You can also think like this:

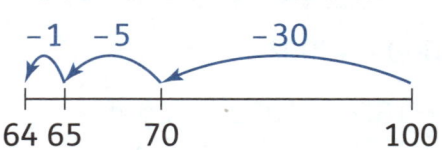

1 Work out the missing number in each number sentence.

a $72 + \boxed{} = 100$ **b** $33 + \boxed{} = 100$ **c** $45 + \boxed{} = 100$

d $\boxed{} + 19 = 100$ **e** $64 + \boxed{} = 100$ **f** $\boxed{} + 61 = 100$

2 Work out how many kilograms you would need to add to each pile of rocks to make 100 kilograms.

57 kg   39 kg   87 kg   13 kg

💡 **Problem solving**

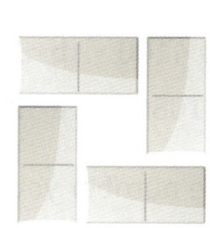

Make the cards and use them to model the solution.

3 How can you arrange these four number cards in a square like the one shown so that the numbers on each side add up to 100?

| 10 | 30 | 60 | 30 | 10 | 50 | 30 | 40 |

➡ *Workbook page 36*

Make 1000

You can use pairs of numbers that add up to 10 or to 100 to find pairs of numbers that add up to 1000.

$1 + 9 = 10$ $\qquad$ $10 + 90 = 100$ $\qquad$ $100 + 900 = 1000$

$4 + 6 = 10$ $\qquad$ $40 + 60 = 100$ $\qquad$ $400 + 600 = 1000$

You can also *mentally* count on or count back to find any pair of numbers that add up to 1000.

$150 + \boxed{} = 1000$

$150 + 50 = 200$ $\qquad$ $200 + 800 = 1000$

So, $\qquad$ $150 + 850 = 1000$

1 Write the matching pairs that add up to 1000.

450 950 400 350 750 150 700

700 550 100 650 850 300 200

2 There are 1000 millilitres in one litre.
How much more juice is needed to fill each of these 1-litre containers?

a $\qquad$ b $\qquad$ c $\qquad$ d

 150 ml $\qquad$  400 ml $\qquad$ 650 ml $\qquad$  850 ml

Problem solving

Draw a bar model to show the problem.

3 How many years is it till the year 3000?

4 An airline carried 4863 passengers in one week. The next week they carried 5000 passengers. How many more passengers did they carry in the second week?

5 A cricket ground can hold 8500 spectators. 6245 people have already bought tickets. How many more tickets are available?

Add multiples of 10

You know that 2 + 3 = 5. So 20 + 30 = 50

2 tens + 3 tens = 5 tens, which is 50

Like you did with smaller numbers, you can group pairs of tens to make it easier to add.

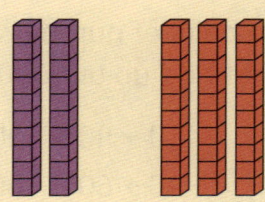

What is 30 + 60 + 70?

3 + 6 + 7	30 + 60 + 70	3 tens + 7 tens = 10 tens, which is 100
= 10 + 6	= 100 + 60	
= 16	= 160	

1 One 10-rod is made of 10 cubes. How many cubes are in each group?

2 Add these numbers as quickly as you can. When you have finished, tell your partner how you worked out the answers.

a 20 + 20 + 20   **b** 30 + 20 + 50   **c** 40 + 80 + 60

d 30 + 60 + 40   **e** 90 + 40 + 10   **f** 40 + 80 + 20

3 Try to do these sums mentally.

a 30 + 40 + 20   **b** 40 + 30 + 30   **c** 30 + 30 + 30

d 40 + 10 + 50   **e** 50 + 10 + 20   **f** 20 + 30 + 30

Problem solving

4 A school bus makes three stops. At the first stop 10 pupils get on, at the second stop 30 pupils get on and at the third stop another 10 pupils get on. How many pupils are on the bus when it gets to school?

Mental strategies for adding

Here are some methods that you can use to add.

Count on in steps

46 + 43

46 + 43 = 89

+10   +10   +10   +10   +1+1+1

46   56   66   76   86 87 88 89

Add the nearest 10 or 100 and compensate

53 + 29 = 82 because it is the same as 53 + 30 – 1

299 + 158 = 457 because it is the same as 300 + 158 – 1

Partition and use place value

35 + 42 = 30 + 5 + 40 + 2

This is the same as 30 + 40 + 5 + 2 = 70 + 7 = 77

Break up the numbers to bridge tens

35 + 48 = 30 + 5 + 40 + 8

This is the same as 30 + 40 + 5 + 8

= 70 + 5 + 8

= 70 + 5 + 5 + 3

= 80 + 3 = 83

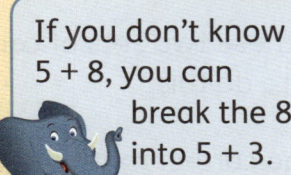

If you don't know 5 + 8, you can break the 8 into 5 + 3.

1 Choose the best strategy to do these additions.

a 47 + 19	**b** 53 + 29	**c** 35 + 39
d 48 + 29	**e** 76 + 39	**f** 88 + 23
g 74 + 58	**h** 23 + 19	**i** 91 + 25

2 Add these numbers using the method you find easiest.

a 42 + 31	**b** 35 + 54	**c** 43 + 16
d 16 + 72	**e** 36 + 61	**f** 57 + 43
g 31 + 82	**h** 19 + 83	**i** 91 + 53

3 What number is 99 more than 43?
Tell your partner how you worked this out.

➡ *Workbook page 37*

Mixed practice 1

1 Look at this set of digit cards.

   a Write the greatest 4-digit number you can make using the cards.

   b What is the smallest 4-digit number you can make using the cards?

   c Copy this number line and show the approximate position of the two numbers you made.

   d Round each number you made to the nearest 100.

2 Cara is making a set of Roman numeral cards. She wants to be able to make all the numbers from 1 to 100.

Which of these two sets of cards will allow her to do this?

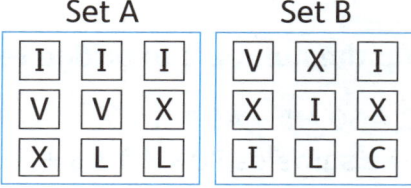

3 What is the correct mathematical name for:

   a a rectangle with four sides the same length

   b a shape with five sides

   c a quadrilateral that has four equal sides but no right angles

   d a regular six-sided polygon?

4 The table shows the times that a bus arrives at five stops. The times are in 24-hour notation.

Stop 132	Stop 133	Stop 134	Stop 135	Stop 136	Stop 137
11:30	11:50	12:05	12:40	12:55	13:00

Where is the bus at these times?

a

b

c

5 **a** Write the decimal numbers shown by the arrows and letters on this number line.

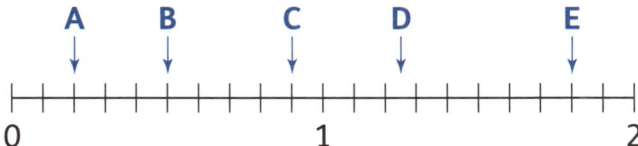

b Write the decimal that is one hundredth greater than each one you wrote in part a.

c What is 24.35 take away 24.3?

6 Sanjita rounded some decimals to the nearest one but she made some mistakes. Find the mistakes and correct them.

a 21.3 → 21   **b** 16.5 → 17   **c** 12.7 → 130   **d** 27.9 → 27

7 **a** Write the mass shown by each arrow on this scale.

b Round each mass to the nearest whole kilogram.

c What is the **difference** between the greatest and the smallest mass shown?

d What is **A** + **B** + **C** + **D**?

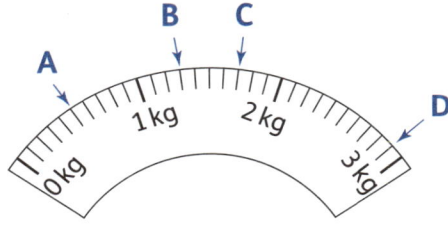

8 A red tank contains 35.5 litres of water. A blue tank contains 35 litres and 450 millilitres of water.

a Which tank contains the most water?

b What is the total amount of water in the two tanks? Give your answer as a decimal.

9 This recipe is for 6 savoury tarts but Emily wants to make 12. How much of each ingredient will she need?

250 g flour
50 ml oil
125 g cheese
4 small tomatoes
200 g chopped onions
0.5 g chilli powder

10 Marwan pays for groceries with a £10 note and three £2 coins.
He receives £1.65 change. How much did his groceries cost?

11 Write the next three numbers in each of these sequences.

a 72 66 60 54   **b** 45 55 65 75

c 99 108 117 126   **d** 350 375 400 425

Symmetry

Is it symmetrical?

Think and share

A shape is **symmetrical** if one half looks the same as the other, but it faces the opposite direction.

This shape is symmetrical.

Explain why each of these shapes is not symmetrical.

Look around the classroom. Find three examples of **symmetry**. Show them to your partner.

1 Which of these shapes are symmetrical? Write the letters in your book.

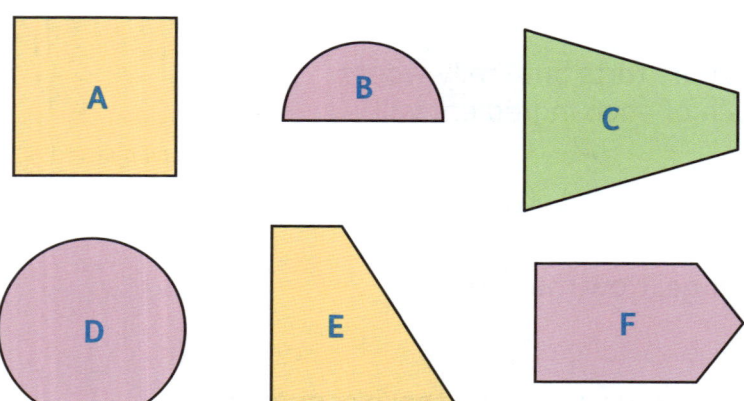

2 Is your face perfectly symmetrical? Use a mirror to examine it closely before you decide.

The dashed line on this shape is a **line of symmetry**.

A line of symmetry is a line that divides a shape into two identical parts.

If you fold the shape along the line of symmetry, the two halves will fit exactly onto each other. We say that each half is the mirror image of the other.

1 Work in pairs. For each shape, decide whether the dashed line is a line of symmetry or not. Give reasons for your answers.

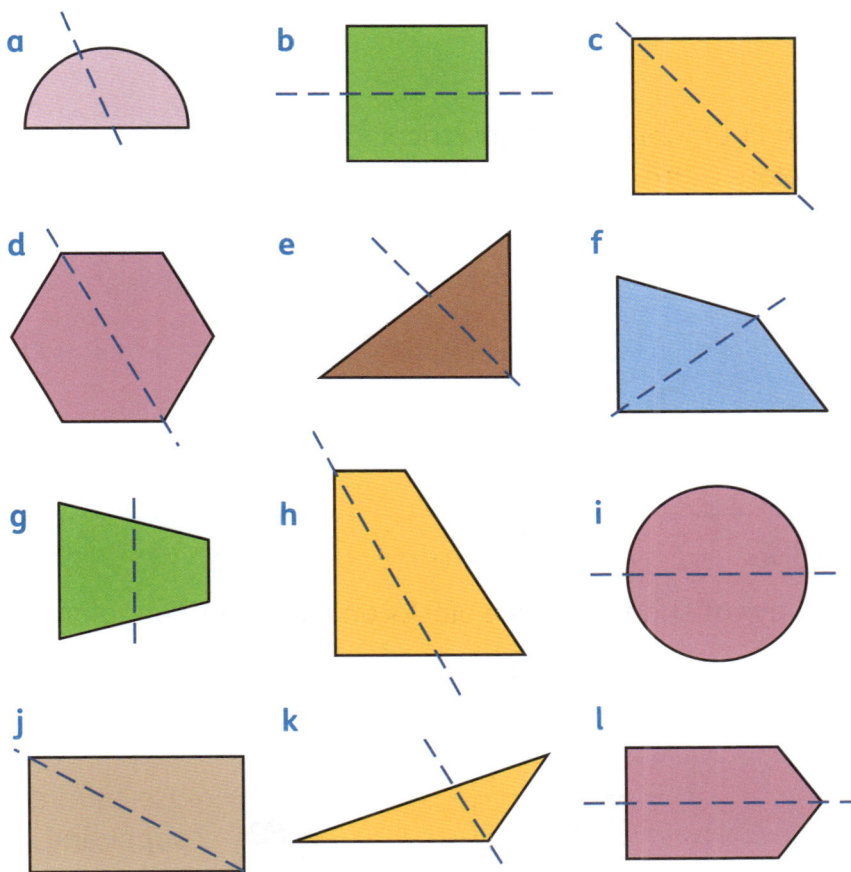

a b c

d e f

g h i

j k l

2 Lines of symmetry are also called mirror lines.
Can you explain why?

Symmetrical shapes

1 Half of a symmetrical shape is given in the first column.
Which is the correct mirror image? Choose from A, B or C.

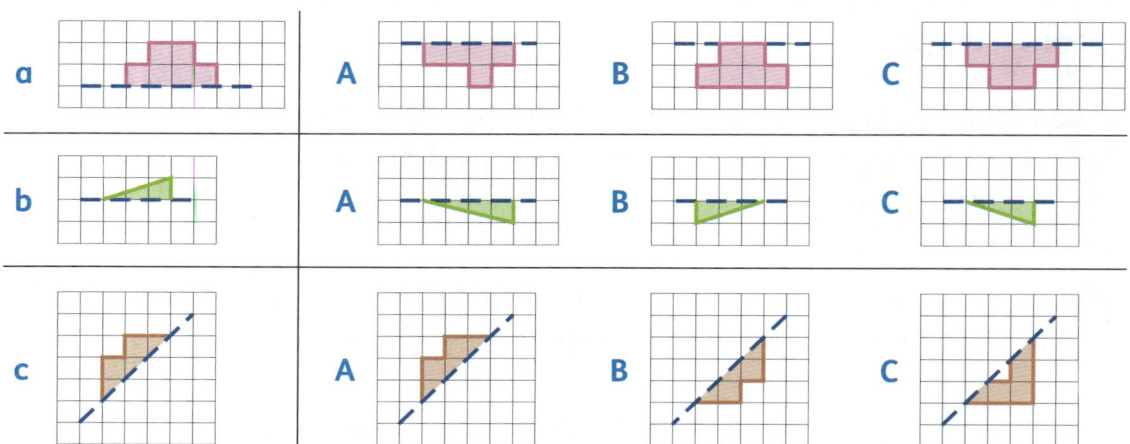

2 These letters and numbers are symmetrical. Draw them in your
book and complete them by drawing the other half of each.

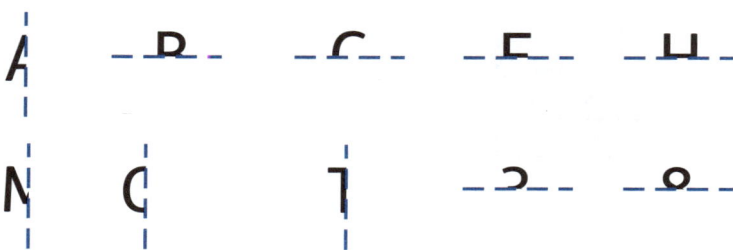

💡 **Problem solving**

Look at these two ways of shading whole blocks on a 2 × 3 rectangular grid.

This shading gives a
symmetrical result.

This shading does not give
a symmetrical result.

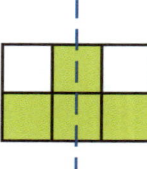
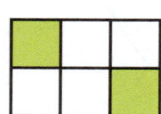

Use squared
paper and work
systematically.

3 How many ways can you find to shade blocks on a 2 × 3 rectangular
grid so that the result has at least one line of symmetry?

➡ *Workbook page 38*

Investigating lines of symmetry

Shapes can have more than one line of symmetry.
The H-shape has a **horizontal** and a vertical line
of symmetry.

Shapes can also have diagonal lines of symmetry.
The square has four lines of symmetry.

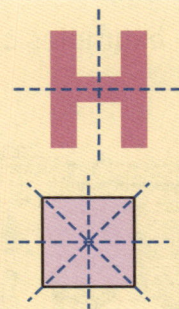

1 Here are five shapes.

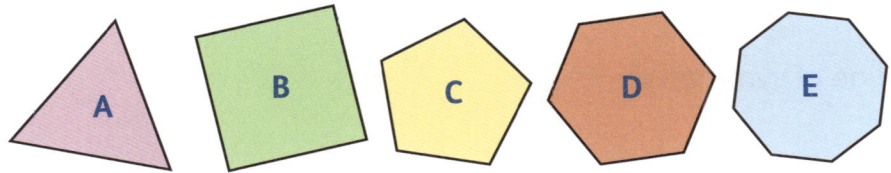

   a Write the name of each shape.

   b Say whether it is symmetrical or not.

   c Draw each shape in your book. Draw the lines of symmetry.
Use a different colour for each line of symmetry.

   d Write the number of sides and the number of lines of
symmetry for each shape. What do you notice?

2 **a** Draw these quadrilaterals on dotted paper.

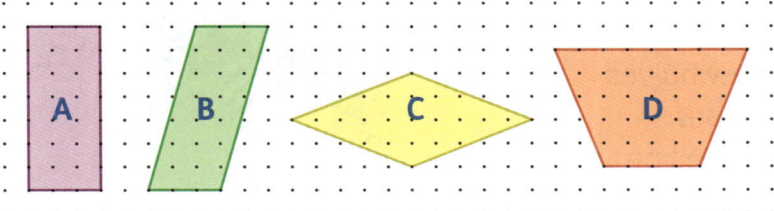

   b Write the name of each shape.

   c If the shape has lines of symmetry, draw them in. Use a
different colour for each line.

   d Why don't these quadrilaterals have four lines of symmetry
like a square?

➡ *Workbook page 39 and page 40*

Symmetry around us

There are many examples of symmetry in daily life.

This building has been designed to be symmetrical.

Can you find the line of symmetry?

1 Find a symmetrical building in your community. Sketch or photograph it and draw the lines of symmetry.

2 This is a rangoli pattern, a traditional symmetrical design from India.

a What shapes can you find in the pattern?

b How many lines of symmetry can you find?

3 These pictures show halves of symmetrical shapes. The dashed line is the line of symmetry.
Work out how many sides the completed shapes would have and name them.

a  b  c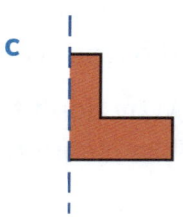

➡ *Workbook page 41*

UNIT 9

Data and charts

Tally charts

Think and share

A **tally** is a way to keep track of items that you count.
If you counted three items, you would make three **tally marks** like this: |||

Each time you count five, you draw the tally mark across the previous four marks like this: ⦀⦀

At the post office, a clerk keeps track of how many people are served each hour like this:

Time	Tally
9 a.m.	⦀⦀ ⦀⦀ |||
10 a.m.	⦀⦀ ⦀⦀ ⦀⦀ ⦀⦀ ||
11 a.m.	⦀⦀ ||||

- Why do you think the clerk uses tallies rather than numbers?

- Which of these sets of tallies is easier to count? Why?

1 A teacher asked pupils to choose two fruits.
The tally chart shows what the pupils chose.

Fruit	Tally
Apples	⦀⦀ ⦀⦀ ⦀⦀ ||||
Pears	⦀⦀ ⦀⦀ |||
Oranges	⦀⦀ ⦀⦀ ||||
Bananas	⦀⦀ ⦀⦀
Peaches	⦀⦀ ||||
Mangoes	⦀⦀ ||

a List the fruits and write how many pupils chose each type.

b Use the chart to work out how many pupils are in the class.

> Remember that each pupil chose two fruits.

2 a Choose six fruits that pupils in your class like.

b Ask ten pupils to choose two fruits.

c Make a tally chart to record the data.

Frequency tables

The **frequency** of a value is the number of times it occurs.

We can use a **frequency table** to show how many times each value occurs.

The frequency is the total of the tallies.

Colour	Tally	Frequency
Black	卌 卌 ‖	12
White	卌 ‖‖	9

1 20 students got these marks out of 10 in a test.

7	6	6	7	5	8	10	9	9	7
7	7	6	5	7	10	8	7	6	7

Draw a frequency table for this data.

2 Look at this frequency table.

Number of books read in the holiday	0–2	3–5	6–8	9 or more
Number of pupils	6	11	3	2

a Work with a partner to make up five questions about the frequency table.

b Swap questions with another pair and answer each other's questions.

💡 Problem solving

3 25 pupils were each asked to choose one of these colours.

Red   **Yellow**   **Blue**   **Brown**   **Green**

- Red was the most popular choice. 11 pupils chose red.
- Only 2 pupils chose yellow.
- Three times as many pupils chose blue than yellow.
- Twice as many pupils chose brown than green.

Work out how many pupils chose each colour. Draw a frequency table to show the data.

➡ *Workbook page 42*

Bar charts

Look at this example to see some of the important features of **bar charts**.

This scale has numbers and units of measure. You use it to find out what amount is shown by each bar.

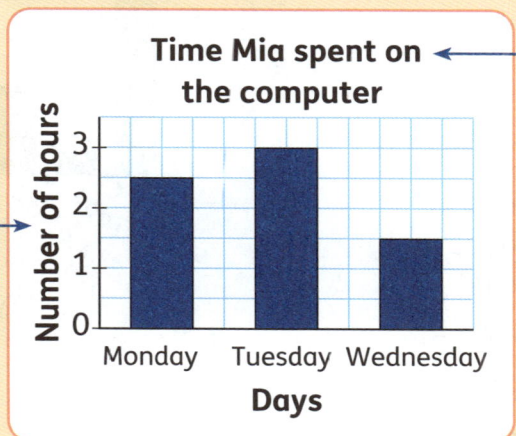

Time Mia spent on the computer

Number of hours / Days
Monday Tuesday Wednesday

The heading tells you what the chart shows.

Bars are the same **width** and the same distance apart.

This scale shows what each bar represents.

1. These statements about bar charts are all false. Explain why.

   a If you label the scales, your graph doesn't need a heading.

   b The number scale should always go up in ones.

   c The width of the bars tells you what amount they show.

2. Write down five pieces of information that you can read from this bar chart.
   For example: *The most common type of litter is chip packets.*

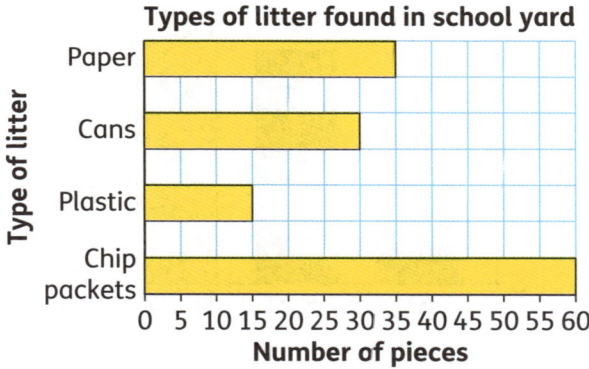

Types of litter found in school yard

Type of litter: Paper, Cans, Plastic, Chip packets
0 5 10 15 20 25 30 35 40 45 50 55 60
Number of pieces

3. This is a very poor example of a bar chart. Discuss what is wrong with it.

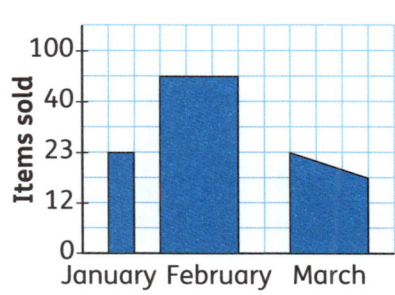

Items sold: 100, 40, 23, 12, 0
January February March

➡ *Workbook page 43*

Use bar charts to answer questions

1 Use the bar chart to answer the questions.

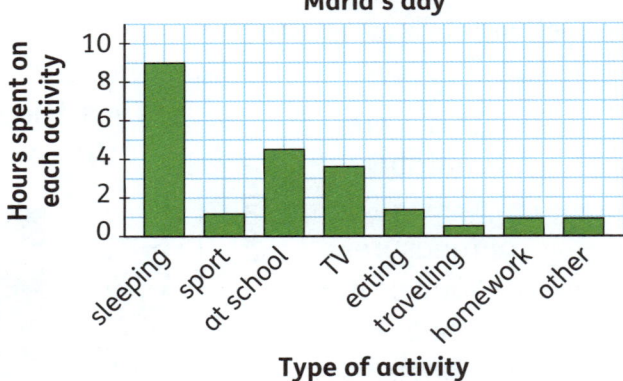

Maria's day

a How long did Maria spend eating?

b How long did she watch TV?

c How much time did she spend altogether at school and doing homework?

d What did she spend the longest time doing?

e What did she spend the shortest time doing?

2 Draw your own bar chart to show how much time you spend doing different activities during the day.

 Problem solving

Read the chart before you start.

3 Look at this bar chart.

These are the answers to five questions about the graph.

Write down what the questions could have been.

a Josh

b Maria

c 9 more than Josh

d 9 fewer than Maria

e 60

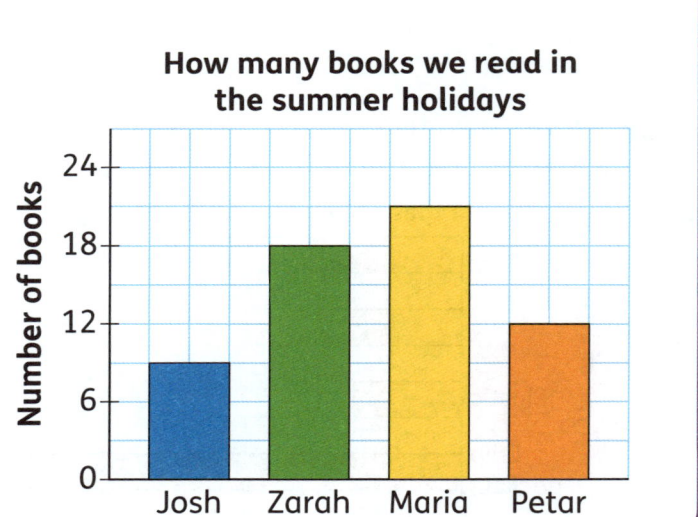

How many books we read in the summer holidays

More bar charts

The coach kept a tally of how many goals the strikers scored in football matches in one season. He asked James and Spike to each draw a bar chart to compare the results.

Name	Tally
Sipho	卌 卌 Ⅰ
Jabu	卌 卌 ‖
James	卌 卌 卌 ‖‖
Rashid	卌 卌 卌 卌 卌 Ⅰ
Solomon	卌 卌 卌 卌
Spike	卌 卌 ‖‖

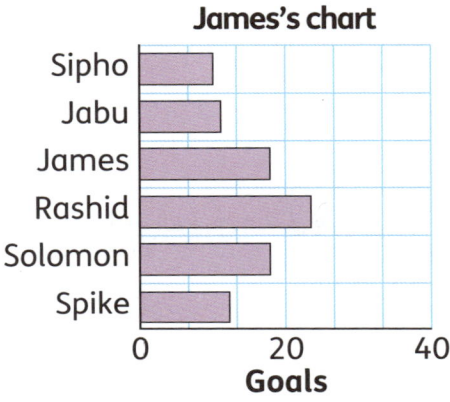

1 Compare the two bar charts.

   a Are both bar charts correct? Give reasons for your answer.

   b Which bar chart do you think is clearer? Why?

   c What makes the bar charts look different?

2 This table shows how much money a taxi driver spent on petrol each month for five months.

September	October	November	December	January
$170	$130	$90	$110	$90

Draw a bar chart to show this information. Use a scale of 1 cm for $20 on the vertical **axis**.

3 Describe how the bar chart would look different if the scale was:

   a 1 cm for $50   b 1 cm for $5

➡ *Workbook page 44*

Pictograms

A **pictogram** is a chart that uses symbols (pictures) to show information. All pictograms should have a clear heading and a **key** that tells you what each symbol represents.

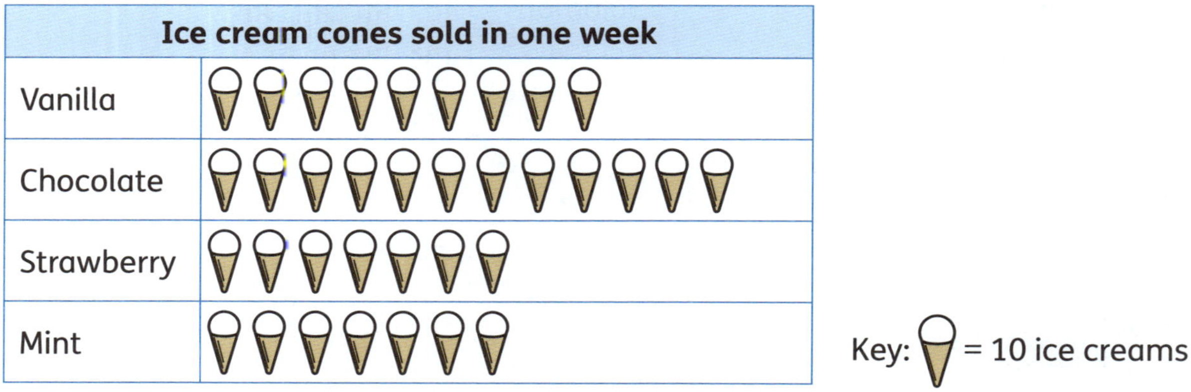

Ice cream cones sold in one week	
Vanilla	♦♦♦♦♦♦♦♦
Chocolate	♦♦♦♦♦♦♦♦♦♦♦
Strawberry	♦♦♦♦♦♦
Mint	♦♦♦♦♦♦

Key: ♦ = 10 ice creams

1 Look at the pictogram and answer the questions.

   a Which was the most popular flavour?

   b Which two flavours were equally popular?

   c How many more vanilla cones were sold than strawberry?

   d How many more chocolate cones were sold than mint?

   e Do you think the pictogram shows the exact number of ice cream cones sold?
   Give a reason for your answer.

   f How could a pictogram like this help someone running an ice cream shop?

2 This tally chart shows the number of ice creams of each flavour sold one weekend.

Vanilla	ЖЖ ЖЖ ЖЖ ЖЖ ЖЖ ЖЖ ЖЖ ЖЖ ЖЖ ЖЖ ЖЖ ЖЖ
Chocolate	ЖЖ ЖЖ ЖЖ ЖЖ ЖЖ ЖЖ ЖЖ ЖЖ ЖЖ
Strawberry	ЖЖ ЖЖ ЖЖ ЖЖ ЖЖ ЖЖ ЖЖ ЖЖ
Mint	ЖЖ ЖЖ ЖЖ ЖЖ ЖЖ
Mango	ЖЖ ЖЖ ЖЖ ЖЖ ЖЖ ЖЖ ЖЖ ЖЖ ЖЖ ЖЖ

Give your pictogram a title and remember to make a key.

Draw a pictogram for this data.

More pictograms

1 This pictogram shows the number of ships arriving at a small harbour on different days of the week.

Number of ships arriving	
Monday	🚢 🚢 🚢 🚢
Tuesday	🚢 🚢
Wednesday	🚢 🚢 🚢
Thursday	🚢 🚢
Friday	🚢 🚢 🚢 🚢 🚢
Saturday	🚢 🚢 🚢 🚢 🚢 🚢 🚢
Sunday	🚢

Key: 🚢 = 5 ships

a On which day did most ships arrive?

b On which two days did the same number of ships arrive?

c How many ships arrived on Monday?

d How many more ships arrived on Saturday than on Friday?

e How many ships arrived altogether this week?

2 Copy and complete this frequency table using information from the pictogram in question 1.

Days of the week	Frequency
Monday–Thursday	
Friday	
Saturday	
Sunday	

3 This pictogram shows the number of different animals that tourists saw on a game drive.

Animals we saw on our game drive	
Lions	◖
Antelopes	⊕ ⊕ ⊕ ⊕ ◖
Giraffes	⊕ ◗
Elephants	⊕ ◕

Key: ⊕ = 20 animals

a Draw the symbols you would use to show:
  • 10 animals
  • 5 animals
  • 15 animals.

b How many different types of animals did the tourists see?

c How many antelope did they see?

d Which animal did they see five of?

e How many more giraffes did they see than lions?

➡ *Workbook page 45*

Venn diagrams

Molly sorted shapes according to two properties: blue shapes and right angles. She used a **Venn diagram** to sort the shapes.

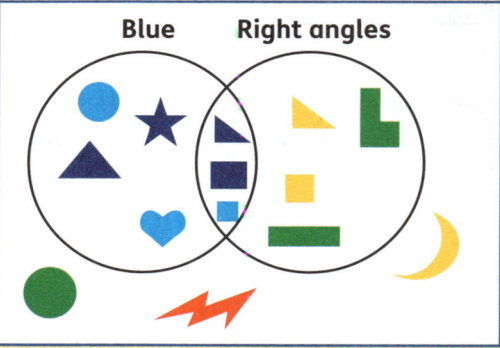

The left-hand circle contains the blue shapes.

The right-hand circle contains shapes with right angles. Some shapes are both blue and right-angled. These go into the overlapping section to show that they belong in both circles.

Some shapes do not belong in either circle. They are placed outside the circles.

1. Draw Venn diagrams to show each pair of sets of numbers. Use all the numbers from 0 to 20 (including 0 and 20) in each Venn diagram.

   a even numbers   numbers > 15

   b even numbers   multiples of 3

   c multiples of 4   numbers < 12

   d odd numbers   numbers > 11

2. These two Venn diagrams are *incorrect*. Find the mistakes and redraw the Venn diagrams correctly.

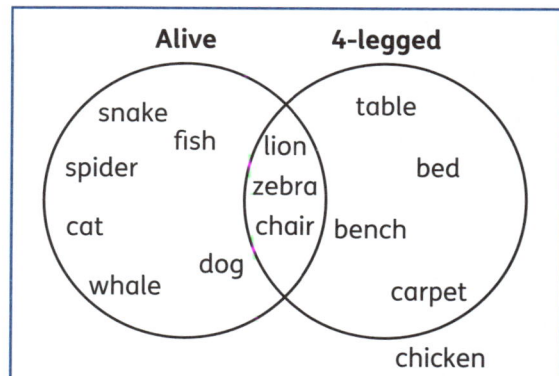

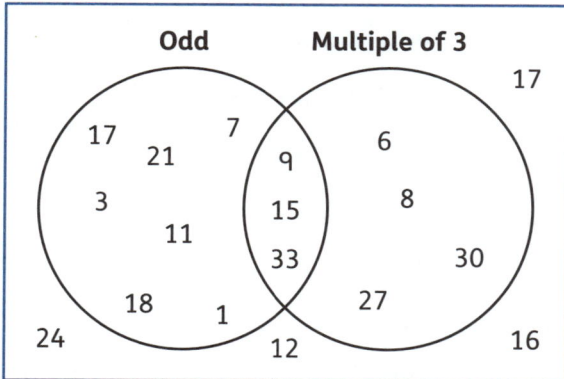

➡ *Workbook page 46*

Solve problems using charts

1 Children from two schools took turns to count the birds in a park. They drew charts to show their data.

Bird count at park (3 August)

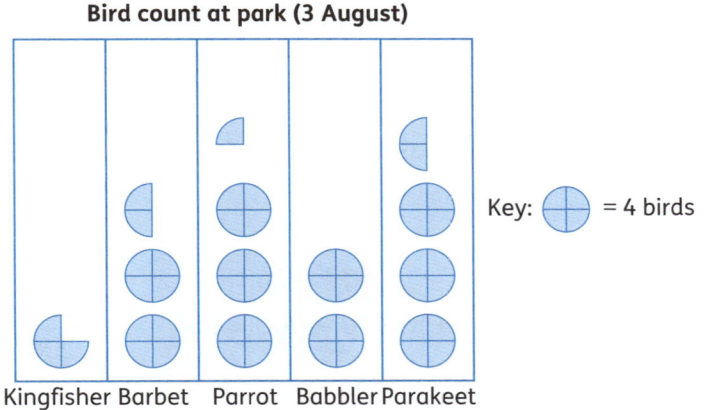

Key: = 4 birds

Kingfisher Barbet Parrot Babbler Parakeet

Bird count at park (22 July)

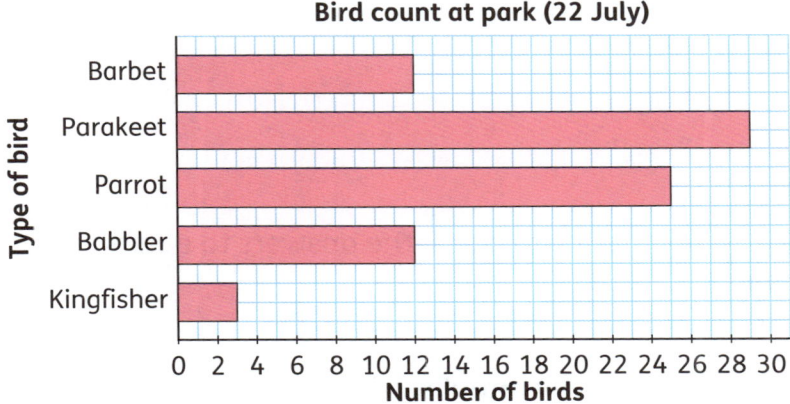

a Compare the charts with a partner. What is similar? What is different?

b Why do you think the groups counted different numbers of birds? Give two reasons.

c What is the difference between the number of babblers counted by each group?

d How many birds did the group count altogether on 3 August?

e Is it reasonable to conclude that parakeets are the most common birds in the gardens? Give a reason for your answer.

f How many more parrots did each group count than kingfishers?

g How many parrots and parakeets were counted on 22 July?

2 Choose one chart on this page. Make up two word problems about the chart. Swap with a partner and solve the problems.

➡ *Workbook page 47*

UNIT 10 Addition and subtraction

Estimate and check

Think and share

- Look at Zara's homework. Explain how she has used rounding to estimate before doing these calculations.

a $273 + 125$

Est: $270 + 130 = 400$

$$\begin{array}{r} 273 \\ + 125 \\ \hline 398 \end{array}$$

b $417 - 209$

Est: $400 - 200 = 200$

$$\begin{array}{r} 4\overset{1}{1}7 \\ - 209 \\ \hline 208 \end{array}$$

Addition undoes subtraction and subtraction undoes addition because they are **inverse operations**.

> $3333 - 1111 = 2222$
>
> So, $1111 + 2222 = 3333$

- How can you use inverse operations to check the answers to addition and subtraction calculations?

1. Calculate the following. Estimate by rounding first. Check your answer using the inverse operation.

   a $87 + 94$ b $43 + 67$ c $99 - 75$ d $76 - 32$

2. Write two subtraction facts for each addition sentence.

   a $321 + 113 = 434$ b $461 + 204 = 665$ c $817 + 349 = 1166$

3. a Draw a number line to show that $321 + 113 = 113 + 321$.

   b Is $876 - 200 = 200 - 876$? Explain your answer.

4. Work with a partner. Discuss how you could use rounding to help you match pairs of numbers that total 500. Try your idea and write the pairs that match.

Explore adding and subtracting larger numbers

1 Three pupils used coloured pens and place-value tables to model adding numbers with three digits.

208 + 391   207 + 383   471 + 249

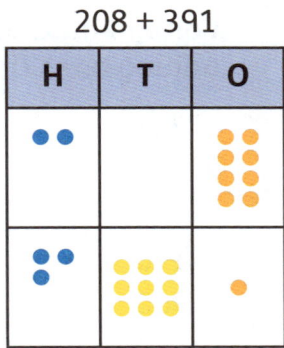 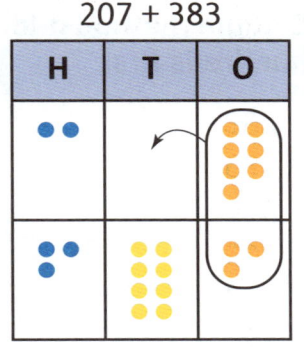 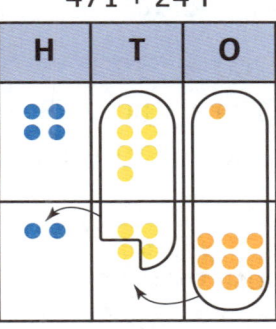

Work in groups. Discuss how the models show addition.
Then answer these questions.

a How do you know which number to draw in the table?

b What do you do if you add digits in a place and you get a 2-digit answer? Why?

c How do you line up the digits when you write the calculations?

2 Write each addition in your book. Calculate the answer.

3 These are their models for subtraction.

435 − 214   606 − 443   863 − 492

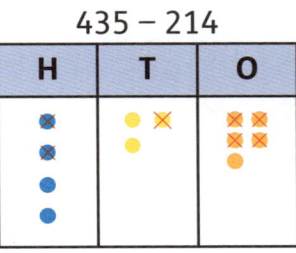 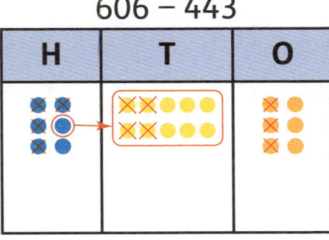

Work in groups. Discuss how the models for subtraction are different from the ones the three pupils used for addition.
Then answer these questions.

a How do you know which number to put at the top of the grid?

b Which digits do you start with, the hundreds or the ones?

c What do you do if the digit being taken away is greater than the one you are taking it from? Why?

4 Write each subtraction in your book. Calculate the answer.

➡ *Workbook page 48*

Use written methods to add

Here are three written methods you can use to calculate 238 + 185

Using place value

$$238 = 200 + 30 + 8$$
$$\underline{+\ 185 = 100 + 80 + 5}$$
$$423 = 300 + 110 + 13$$

Which of these methods
do you like best? Why?

Column method adding hundreds first

```
   238
 + 185
   300
   110
    13
   423
```

Column method adding ones first

```
   238
 + 185
    13
   110
   300
   423
```

Look at these two column additions.

```
   426
 + 290
   716
 1
```

```
  2473
 + 329
  2802
  1 1
```

Look at the place-value models for addition on page 75 again. What do the numbers below the totals represent in these column additions?

1 Rewrite each calculation as a column addition. Estimate and then work out the answer.

a 231 + 128   b 327 + 184   c 841 + 193

d 305 + 105   e 228 + 336   f 818 + 245

g 1245 + 632   h 342 + 1902   i 754 + 686

💡 Problem solving

2 Look at these two bar models.

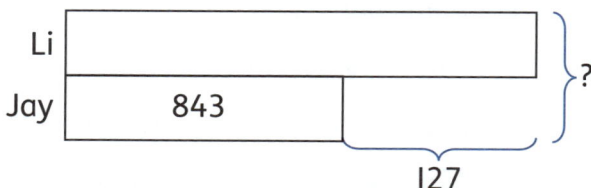

a Make up your own word problem to match each model.

b Calculate the total for each problem.

Use written methods to subtract

Subtraction involves finding the difference between two numbers.

You already know you can count up from the smaller to the greater number to find the difference.

In this example, the number line shows how to calculate 336 – 178 by counting up in steps from 178 to 336.

The column subtraction shows the same calculation.

Counting up in steps using a number line

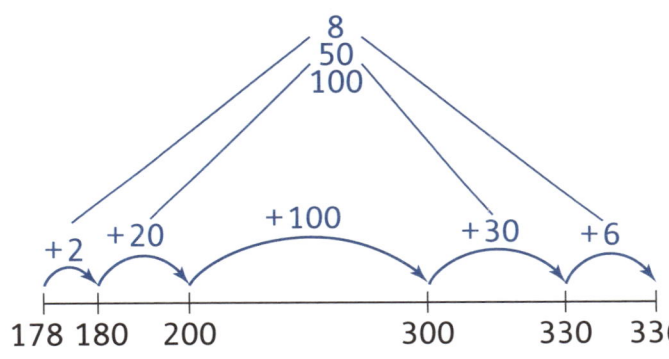

Column method showing totals at the side

```
   336
 - 178
     2  + 180
    20  + 200
   100  + 300
 +  36  + 336
   158
```

You can use what you know about place value to write subtractions in a more efficient way.

The table shows how to get more ones and tens by exchanging tens and hundreds.

In the column subtraction, the exchanges are shown by crossing out and carrying.

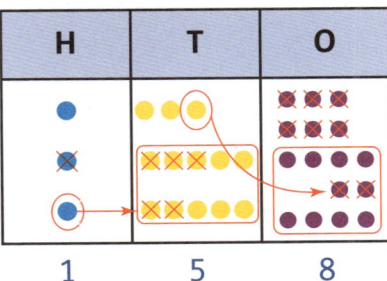

```
   2 12 1
   3̶3̶6
 - 178
   158
```

1. Rewrite each calculation as a column subtraction. Estimate and then work out the answer.

   a 692 – 545   b 321 – 267   c 816 – 797

   d 924 – 799   e 1211 – 158   f 2734 – 686

Subtraction problems

In this code, each letter of the alphabet stands for a digit from 0 to 9:

A	B	C	D	E	F	G	H	I	J	K	L	M
0	1	2	3	4	5	6	7	8	9	0	1	2

N	O	P	Q	R	S	T	U	V	W	X	Y	Z
3	4	5	6	7	8	9	0	1	2	3	4	5

1 Solve these problems using the code.

a Subtract ABC from DEF. Write the number.

b What number is the difference between POD and PEA?

c Write a code for the answer to 471 – 289.

d What number is the difference between HIM and HER?

e Subtract LMN from HIJ. Write the number.

f Write a code for the answer to 689 – 302.

g What number is the difference between FISH and CAT?

h Subtract WXY from STU. Write the number.

i Write a code for the answer to 842 – 377.

j What number is the difference between CAR and BUS?

2 Make up your own coded subtractions.

3 In these subtractions, some of the digits are missing. Copy them and fill in the missing digits. Check they are correct using the inverse operation.

a $439 - 1\square6 = 31\square$

b $5\square7 - 263 = \square74$

c $\square25 - 532 = 1\square3$

d $86\square - \square45 = 623$

💡 Problem solving

4 Look at these two bar models.

945 beads	
473 red	? blue

929 passengers	
? passengers	124 more

a Make up your own word problem to match each model.

b Calculate the missing amounts.

➡ *Workbook page 49*

Do you add or subtract?

Problem solving

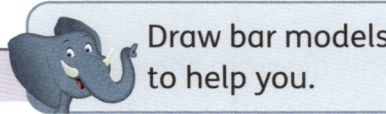
Draw bar models to help you.

1 Answer these questions about the fruits and nuts. Estimate each answer. Check each answer by doing the inverse operation.

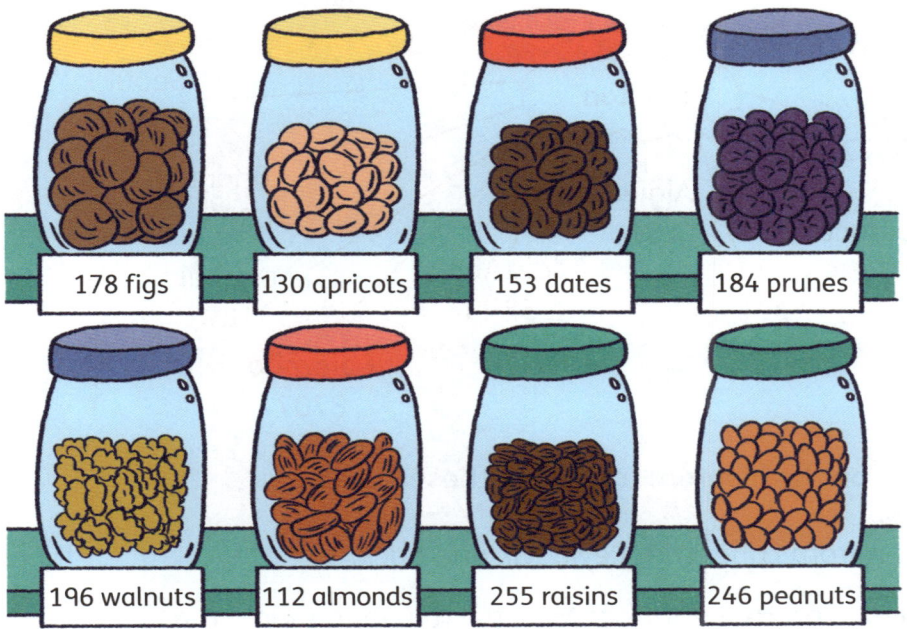

178 figs | 130 apricots | 153 dates | 184 prunes

196 walnuts | 112 almonds | 255 raisins | 246 peanuts

a How many figs and prunes are there altogether?

b How many more walnuts are there than almonds?

c How many apricots and figs are there altogether?

d What is the difference between the number of peanuts and raisins?

e I want to give my friend 200 dates. How many more will I need?

2 Use the numbers in the boxes. What do you need to add to each number to make 950?

246	353	537
883	902	364
518	672	826

3 Maria used a calculator to subtract a number from 2154. The answer was 1478. Maria can't remember what number she subtracted. How could she work it out?

➡ *Workbook page 50*

Practise adding and subtracting

This diagram shows how many kilometres a plane would fly from London to some other cities.

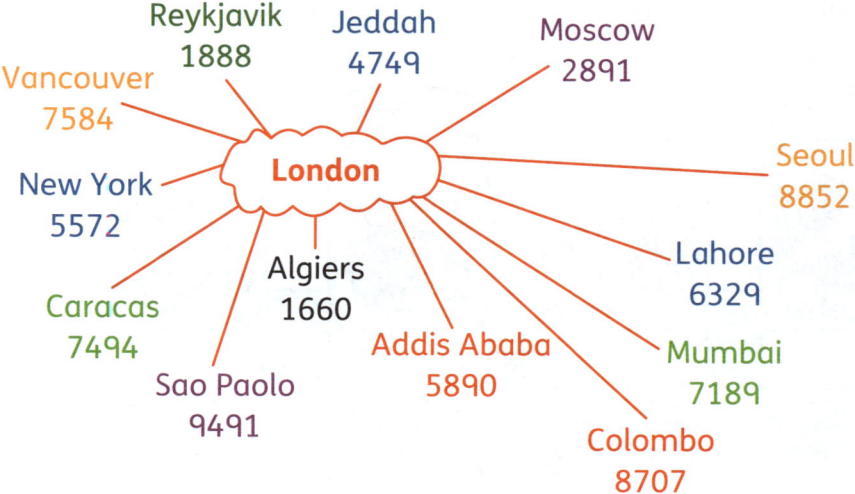

Reykjavik 1888
Jeddah 4749
Moscow 2891
Vancouver 7584
Seoul 8852
New York 5572
London
Lahore 6329
Algiers 1660
Caracas 7494
Addis Ababa 5890
Mumbai 7189
Sao Paolo 9491
Colombo 8707

Use the distances on the diagram to answer these questions.

1 How far is it from London to Reykjavik and back?

2 A pilot does a return flight from Moscow to London. How many kilometres is this in total?

3 How much further is it from London to Mumbai than from London to Lahore?

4 A plane flies from Jeddah to London, then onto Mumbai. How far will it fly from Jeddah to Mumbai?

5 How much closer to London is Algiers than Sao Paolo?

6 Which is closer to London, Colombo or Caracas? How much closer is it?

7 Zayn flew from New York to London and Nia flew from Seoul to London. How much further did Nia fly?

8 Subtract the shortest distance on the diagram from the longest distance.

9 Work out the distance a plane would fly from Moscow to London and then onto Vancouver.

10 Make up an addition question and a subtraction question of your own for this diagram. Exchange with a partner and work out the answers to each other's questions.

Solve two-step problems

A fruit seller bought mangoes, plums and pineapples. She bought 450 plums and 230 more mangoes than plums. Altogether she bought 1475 pieces of fruit. How many pineapples did she buy?

You can draw a bar model to show the information.

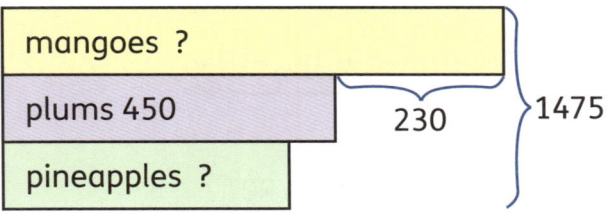

You can work out the number of mangoes.

450 + 230 = 680 mangoes

You can work out the number of mangoes and plums:

450 + 680 = 1130

Once you know the number of mangoes and plums you can subtract this from the total to find the number of pineapples.

> You can also find the combined number of mangoes and plums by adding 450 + 450 + 230.

1475 − 1130 = 345   She bought 345 pineapples.

 Problem solving

 Draw a bar model to help you.

1. Zayne has 127 more beads than Maria. Maria has 512 beads. How many beads do they have together?

2. At lunch time there were 461 cars parked in a lot. During the afternoon, 127 cars drove away and some more arrived. At the end of the day there were 442 cars parked in the lot. How many cars arrived during the afternoon?

3. Two countries have a combined **area** of 8379 square kilometres. Country A is 1087 square kilometres smaller than Country B. What is the area of each country?

Angles and triangles

Types of angles

> **Think and share**
>
> - In maths an angle is a measure of turn. Use these diagrams to explain what that means:
>
>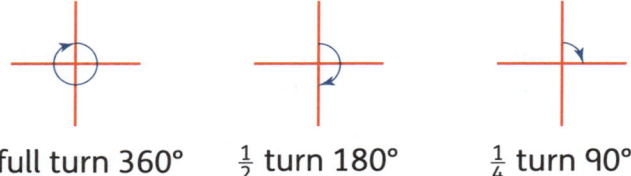
>
> A full turn 360° $\frac{1}{2}$ turn 180° $\frac{1}{4}$ turn 90°
>
> - How can you check whether an angle is a **right angle** or not?
>
>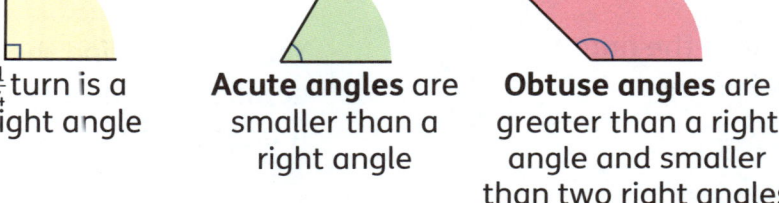
>
> $\frac{1}{4}$ turn is a right angle
>
> **Acute angles** are smaller than a right angle
>
> **Obtuse angles** are greater than a right angle and smaller than two right angles
>
> - Find an example of each type of angle in the classroom. Explain how you decided what type of angle each one was.
>
> Finn says Angle B is greater than Angle A. Mo says they are the same size and that the length of the arms doesn't matter.
>
> Who is correct? How could you convince the other person?
>
>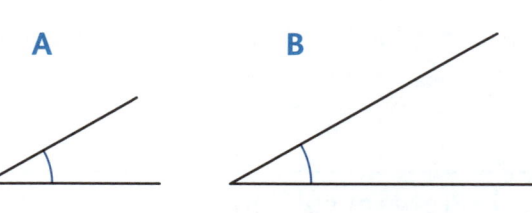

1   What type of angle is formed between the hands on each clock?

a   b   c
d   e   f

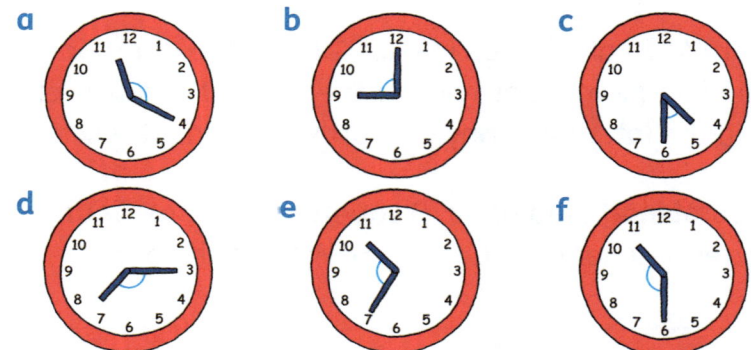

Compare and order angles

1 Angles **A** to **I** are marked on this drawing of a lion.

   a Find the smallest and the greatest angles in the diagram. How could you check you are correct?

   b Circle the acute angles in your list.

   c Underline the obtuse angles.

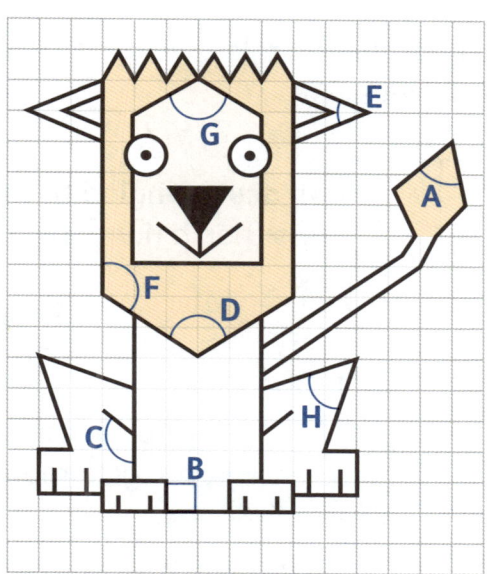

2 Work in pairs. For each marked angle, try to find another angle in the drawing that is the same size.

3 Arrange the three angles in each set in order of size, from greatest to smallest.

a

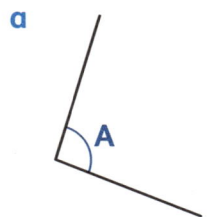  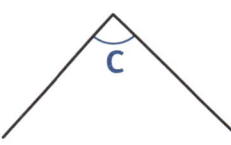

b

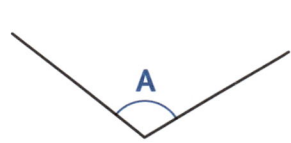  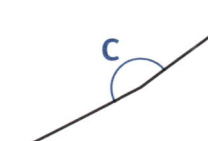

 Problem solving

4 Fahim made this decorated initial for an art project.

   a How many right angles are inside the F?

   b How many of the angles inside the F are acute?

   c Are there any obtuse angles?

➡ *Workbook page 51 and page 52*

Types of triangles

To describe triangles, we can talk about their angles or the lengths of their sides.

 An **acute-angled triangle** has all its angles less than a right angle.

 A **right-angled triangle** has one angle of 90°.

 An **obtuse-angled triangle** has one angle greater than a right angle.

A triangle with no sides the same length is called **scalene**.

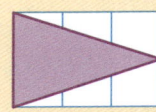 A triangle with two sides the same length is called **isosceles**.

A triangle with three sides the same length is called **equilateral**.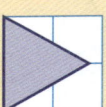

1 For each of these triangles say whether it is acute-angled, right-angled or obtuse-angled.

a 　　b 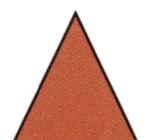　　c 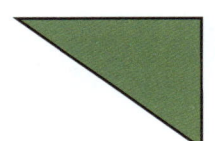　　d

2 For each of these triangles say whether it is scalene, isosceles or equilateral.

a 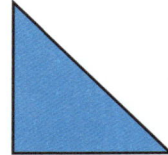　　b 　　c 　　d

Classify triangles

You can group and name triangles using the properties of their angles and their sides. For example, triangle A is right-angled and it has no equal sides. So, triangle A is a right-angled scalene triangle.

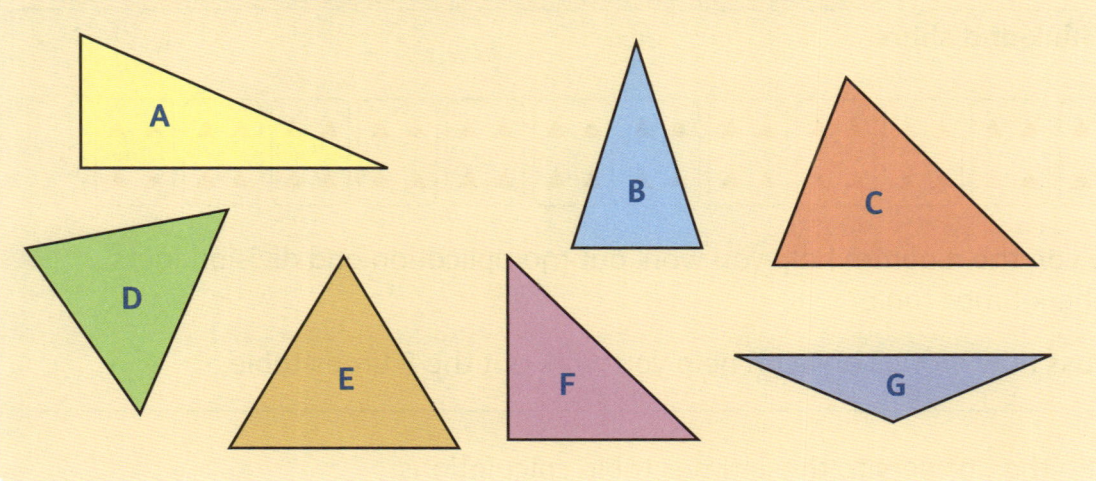

1 **Classify** triangles B to G by saying whether each is:

  a acute-angled, obtuse-angled or right-angled

  b scalene, isosceles or equilateral.

There may be more than one answer to some problems.

💡 **Problem solving**

2 Which triangle is the odd one out in this set?
Give reasons for your answers.

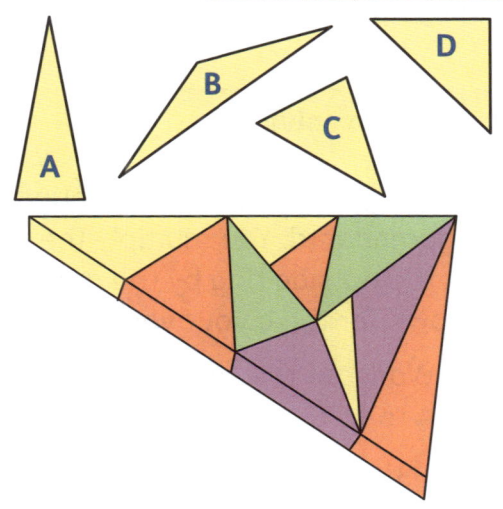

3 Max made this set of painted wooden triangles.
How many scalene triangles are in the set?

➡ *Workbook page 53*

Multiplication and division facts

Revise multiplication facts

💭 Think and share

- How can the diagram help you work out multiplication and division facts for the 4 times table?

- Discuss how groups of 4 can help you work out the 8 times table.

1 Write the answers to these times table calculations.

a 2 × 6	b 2 × 8	c 2 × 9	d 4 × 7
e 4 × 9	f 3 × 2	g 3 × 8	h 3 × 10
i 5 × 7	j 5 × 1	k 10 × 2	l 10 × 10

2 What is the **product** of these pairs of numbers?

a 2 and 6	b 3 and 5	c 10 and 4
d 3 and 7	e 4 and 5	f 10 and 8

> When you multiply numbers, the answer is the product of those numbers.

💡 Problem solving

3 In this game, you throw a counter onto a gameboard. You multiply the number on the outside ring by the amount shown for the colour you land on.

a Work out the scores for counters A to I.

b Li threw two counters and got a score of 72. Where could his counters have landed?

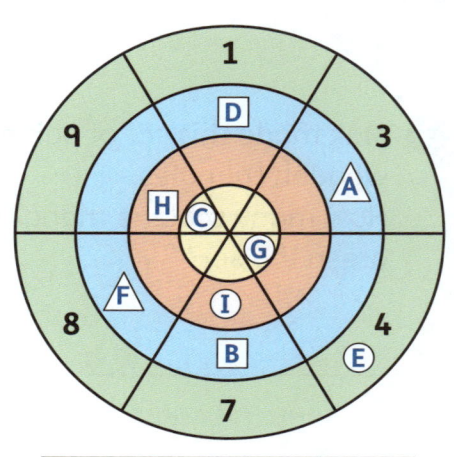

Key: ☐ ×1 ☐ ×8 ☐ ×4 ☐ ×10

More tables

You already know the 2, 3, 4, 5, 8 and 10 times tables. This year you will learn the facts for the 6, 7, 9, 11 and 12 times tables.

Here are the facts for the 6 and 9 times tables. The sets of cubes show groups of 6 and groups of 9.

Remember that you already know many of the facts for the other tables because 3×6 is equal to 6×3 and 4×12 is equal to 12×4 and so on.

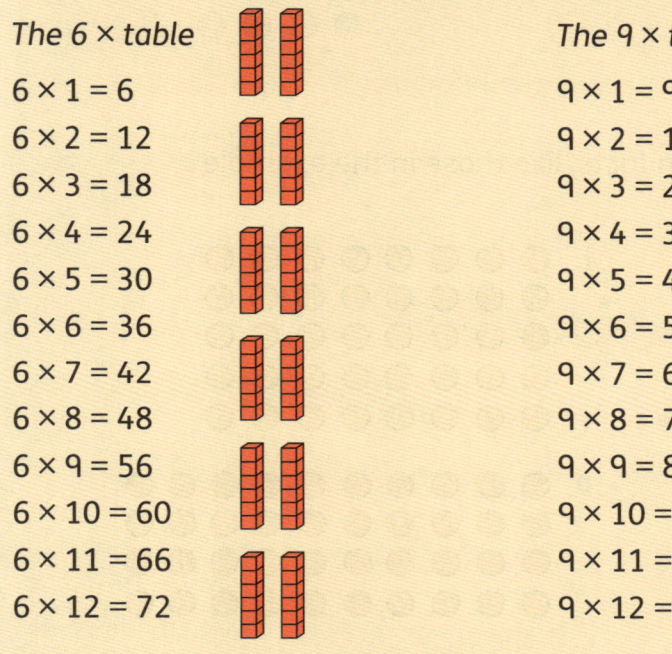

The 6 × table

$6 \times 1 = 6$
$6 \times 2 = 12$
$6 \times 3 = 18$
$6 \times 4 = 24$
$6 \times 5 = 30$
$6 \times 6 = 36$
$6 \times 7 = 42$
$6 \times 8 = 48$
$6 \times 9 = 56$
$6 \times 10 = 60$
$6 \times 11 = 66$
$6 \times 12 = 72$

The 9 × table

$9 \times 1 = 9$
$9 \times 2 = 18$
$9 \times 3 = 27$
$9 \times 4 = 36$
$9 \times 5 = 45$
$9 \times 6 = 54$
$9 \times 7 = 63$
$9 \times 8 = 72$
$9 \times 9 = 81$
$9 \times 10 = 90$
$9 \times 11 = 99$
$9 \times 12 = 108$

It is useful to learn times tables. Knowing multiplication facts will help you to multiply larger numbers. Multiplication facts also help you divide.

1 Work in pairs to do this activity.

a Make a set of small cards by cutting up some stiff paper or card.

b Work with a partner. Test each other to see which multiplication facts you know from any of the times tables up to 12.

c If you don't know a fact, write it on one side of a card.

4×8

d Look up, or work out, the correct answer. Write it on the back of the card.

e Keep the cards with you and keep testing yourself until you know all the facts.

➡ *Workbook page 54*

Arrays

This arrangement of dots is called an **array**.
Look at these facts about the number 24.

6 rows of 4 = 24

4 columns of 6 = 24

$6 \times 4 = 24$ $\qquad$ $4 \times 6 = 24$

$24 \div 4 = 6$ $\qquad$ $24 \div 6 = 4$

1 Write multiplication and division facts, like those in the example above, for each array.

a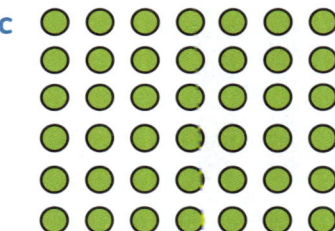

b

c

d

2 Use counters or stones to show each of these numbers as an array.

    a 32   **b** 45   **c** 48   **d** 36

Problem solving

3 Make up two multiplication and two division word problems that can be solved using the arrays in question 1.

4 Swap problems with a partner and use the arrays to find the solutions to your partner's word problems.

Grids

You can use grids drawn on squared paper to help you find the answers to multiplication problems.

Look at these two grids.

This grid shows 2 × 4.

You can count the squares to find the product. 2 × 4 = 8.

You can also use the grid and count squares to find 2 × 1, 2 × 2 and 2 × 3.

This grid shows 3 × 8.

You can count the squares to find the product. 3 × 8 = 24.

You can also use this grid to find other products. For example, 3 × 3 or 6 × 2.

1 Draw a blank 10 × 10 grid of squares like this one. Use it to find each product.

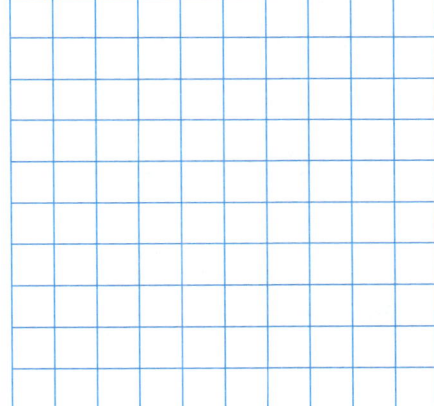

    a 5 × 7   **b** 8 × 8

    c 3 × 9   **d** 6 × 8

    e 4 × 10   **f** 8 × 9

    g 7 × 7   **h** 9 × 5

    i 10 × 6   **j** 7 × 8

2 Explain how you could use the 10 × 10 grid in question 1 to help you work out:

    a 11 × 4

    b 12 × 5

3 How can you use grids of squares to help you solve division problems?
Share your ideas with your group.

Practise your facts

> Work out the value of each letter before you start.

 Problem solving

This grid shows a code.

To use it, you multiply two numbers to find a letter.

$5 \times 4 = 20$

20 is the code for V.

×	6	9	7	4
5	O	W	T	V
8	E	L	M	A
3	U	S	D	I
10	N	C	P	K

1 a Use the grid to decode this message.

| 45 | 48 |

| 72 | 3 × 4 | 40 | 8 × 6 |

| 56 | 18 | 9 × 8 | 35 | 12 | 70 | 72 | 4 × 3 | 90 | 32 | 5 × 7 | 12 | 30 | 60 |

| 8 × 4 | 6 × 10 | 3 × 7 |

| 21 | 12 | 20 | 12 | 3 × 9 | 3 × 4 | 5 × 6 | 60 |

b Use the grid to make your own coded message.

c Ask a partner to decode it.

2 Make your own times table strips game.

You will need: 10 counters and set of blank times table strips.

×	1	2	3	4	5	6	7	8	9	10	11	12
5	5	10	15	20	25	30	35	40	45	50	55	60

Each player chooses a times table and makes a times table strip like the one above, with all the numbers filled in.

> You can choose any times table.

Decide how many numbers on the strip to cover. The first player places counters on their strip and the second player works out which numbers are covered.

The second player wins the counters from the numbers they correctly worked out.

Swap over. Now the other player must try to collect counters.

The player with the most counters wins.

➤ *Workbook page 55*

Patterns and fact families

When we multiply, we multiply **factors** to get a product.

$$5 \times 8 = 40 \qquad\qquad 8 \times 5 = 40$$

factors   product   factors   product

We can multiply factors in any order, so $8 \times 5 = 5 \times 8$.

Inverse operations undo each other.

Division is the inverse of multiplication.

$5 \times 8 = 40$

So, $40 \div 5 = 8$ and $40 \div 8 = 5$

A **fact family** is a set of the related multiplication and division facts for a set of three numbers.

If you know one fact, you can use it to work out all the others.

> What is the fact family for 11, 4 and 44?
>
> $11 \times 4 = 44 \qquad 4 \times 11 = 44 \qquad 44 \div 11 = 4 \qquad 44 \div 4 = 11$

1 Write the fact family for each set of three numbers.

a 3, 6, 18

b 7, 6, 42

c 7, 9, 63

d 8, 4, 32

e 6, 9, 54

f 5, 6, 30

g 7, 8, 56

h 9, 9, 81

i 8, 9, 72

j 12, 6, 72

k 11, 5, 55

l 12, 12, 144

2 a Write the fact family for 5, 5 and 25.

b Why does this fact family only have two facts?

c Can you think of another set of three numbers that will also give only two facts? Explain how you decided.

Problem solving

3 Apples are sold in bags of 5.

a How many apples are there in 12 bags?

b There are 55 apples. How many bags can you fill?

Division facts

You can use multiplication facts to find division facts.

For example, you know that $6 \times 8 = 48$
So, $48 \div 6 = 8$ and $48 \div 8 = 6$

To find division fact $48 \div 6$ ask yourself, 'What times 6 gives me 48?'

This can be written as $\square \times 6 = 48$ or $6 \times \square = 48$

1 Use multiplication facts to find the missing numbers in the division sentences.

a $4 \times 6 = 24$ $24 \div 4 = \square$ $24 \div 6 = \square$

b $5 \times 7 = 35$ $35 \div 5 = \square$ $35 \div 7 = \square$

c $6 \times 7 = 42$ $42 \div 6 = \square$ $42 \div 7 = \square$

d $9 \times 4 = 36$ $36 \div 9 = \square$ $36 \div 4 = \square$

e $10 \times 8 = 80$ $80 \div 10 = \square$ $80 \div 8 = \square$

2 Complete these number sentences.

a $\square \times 5 = 40$ b $\square \times 3 = 27$ c $\square \times 4 = 16$

d $\square \times 2 = 18$ e $\square \times 6 = 42$ f $\square \times 7 = 21$

g $5 \times \square = 25$ h $8 \times \square = 56$ i $6 \times \square = 54$

j $4 \times \square = 28$ k $6 \times \square = 48$ l $10 \times \square = 100$

3 Write the answers to these divisions.

a $30 \div 6 = \square$ b $35 \div 7 = \square$ c $63 \div 9 = \square$

d $24 \div 6 = \square$ e $90 \div 9 = \square$ f $30 \div 5 = \square$

g $21 \div 3 = \square$ h $45 \div 5 = \square$ i $16 \div 4 = \square$

j $27 \div 3 = \square$ k $56 \div 8 = \square$ l $81 \div 9 = \square$

➡ *Workbook page 56*

Multiply and divide by 1 and 0

Why do you think we don't need a 1 times table?

$1 \times 5 = 5$ $1 \times 6 = 6$ $1 \times 8 = 8$ $1 \times 12 = 12$

There is a rule for multiplying any number by 1.

The product of any number n and 1 is the number itself.

$n \times 1 = n$

When you divide a number by 1, the answer is the number itself.

$n \div 1 = n$

Look at the flowers again. Divide the number of petals by 1.

Now think about flowers with no petals.

> Remember your fact families. $1 \times 5 = 5$, so $5 \div 1 = 5$. This gives us a rule for dividing any number by 1.

$1 \times 0 = 0$ $2 \times 0 = 0$ $3 \times 0 = 0$

Can you see a pattern?

The product of any number and 0 is always 0.

$n \times 0 = 0$

Suppose that n friends share nothing. Each person's share is nothing!

So, 0 divided by any number n is 0.

$0 \div n = 0$

In maths we don't divide by 0. We say that dividing by 0 is meaningless because if you divide no times, then you are not actually dividing.

1   Read each statement. Say whether it is true or false.

    a $346 \times 1 = 1$ **b** $346 \div 1 = 346$ **c** $45 \times 0 = 0$ **d** $230 \times 1 = 230$

More properties of multiplication

What do these two arrays tell you about multiplication?

You can change the order of numbers in multiplication and you will still get the same answer. For example,
$12 \times 4 = 48$ and $4 \times 12 = 48$

This property of multiplication means you can change the order of numbers to multiply.

$2 \times 3 = 6$

$3 \times 2 = 6$

What is $4 \times 2 \times 5$?

Multiply	This is the same as
$4 \times 2 \times 5$	$2 \times 5 \times 4$
$= 8 \times 5$	$= 10 \times 4$
$= 40$	$= 40$

Multiply $12 \times 4 \times 2$ $12 \times 4 \times 2$
$= 48 \times 2$ $= 12 \times 8$
$= ?$ $= 96$ $(12 \times \text{table})$

Always look at the numbers and think about the order that is easier for you.

1 Copy and complete these number sentences.

a $12 \times 8 = 8 \times \square$

b $19 \times \square = 3 \times 19$

c $2 \times 4 \times 3 = \square \times 3$

d $13 \times 19 = 19 \times \square$

e $3 \times 12 \times 2 = 2 \times 3 \times \square$

f $14 \times 0 = 0 \times \square$

g $1 \times 35 = 35 \times \square$

h $1 \times 2 \times 3 = 2 \times \square \times \square$

i $2 \times 2 \times 3 = \square \times 3$

2 Hisham says:

When you have to multiply three numbers, you have to work from left to right.

How would you show him that this is not true?

1 Read these statements and make up an example to show what each pupil means.

a

> When you have to multiply two numbers, you can do the calculation in any order.

b

> It doesn't matter how many numbers you have to multiply – if you see × 0 in the calculation, you can just write 0 as the answer.

c

> You can multiply in any order, but you can't change the order when you have to divide.

d

> I had to multiply a number by 12 but I forgot my 12 times table so I multiplied by 2 and then I multiplied by 6.

e

> When you have to multiply a few numbers, you can do the calculation in any order to make it easier.

Problem solving

There is more than one possible answer for some of these questions.

2 Use the information given for each set of cards to work out what numbers the letters could represent. Write the letters and the values in your book.

a $x \times 8 = 0$

b $p \times q \times r = 7 \times 6$

c $p \times q \times r = 24$

d $a \times b \times c = a + b + c = a \times b \times c$

e $4 \times 4 \times 3 = 2 \times 8 \times d = e \times f \times 12$

➡ *Workbook page 57*

UNIT 13 Negative numbers

Numbers less than 0

Think and share

Look at the flcor numbers on the elevator. What do they tell you?

Numbers that are less than 0 are called **negative numbers**.

- Think of some other examples of negative numbers in real life.

We extend the number line to the left of 0 to show negative numbers.

0 itself is neither positive nor negative.

We use a **minus sign** to show that a number is less than zero.

−10 −9 −8 −7 −6 −5 −4 −3 −2 −1 0 1 2 3 4 5 6 7 8 9 10

- Is −10 greater than or less than −9? Explain your answer.

1. Write the missing numbers from each number line.

a
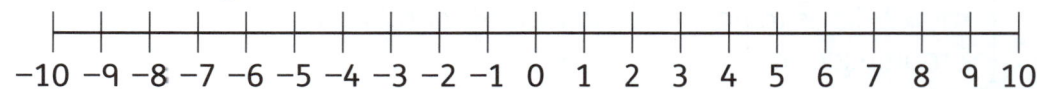
−6 ☐ −4 ☐ ☐ −1 0 1 2 ☐ 4 ☐ 6 7 8

b −25 −24 ☐ ☐ −21 −20 −19
c −150 −100 −50 ☐ ☐

d
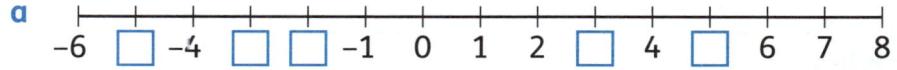
−60 ☐ ☐ ☐ 20 40 ☐ ☐

e ☐ ☐ ☐ 0 10 ☐ 30

2. Copy and fill in <, > or = between each pair of numbers.

a 25 ☐ −25   b −5 ☐ −6   c 0 ☐ −5

d −10 ☐ −5   e −2 ☐ −1   f −100 ☐ −200

3. How did you think and work to compare the numbers in question 2?
   Write a tip to help people compare negative numbers.

➡ *Workbook page 58*

Negative temperatures

We measure temperature in **degrees Celsius** using a thermometer.

°C means 'degrees Celsius'. 0 °C is the temperature at which water freezes (turns to ice). Negative numbers on a thermometer show us temperatures below freezing.

You read the scale on a thermometer like a number line.

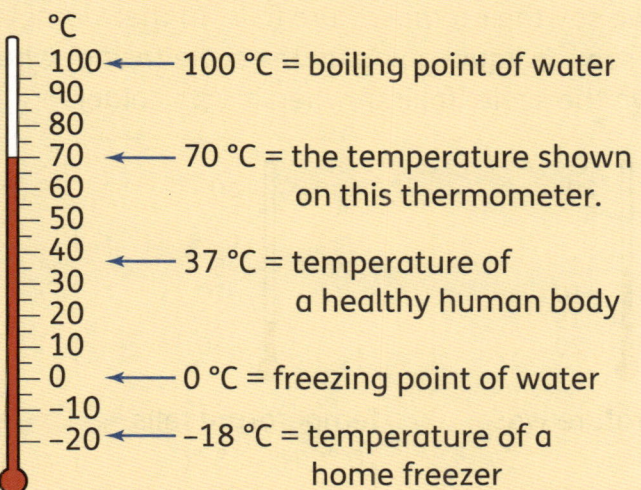

°C
- 100 ← 100 °C = boiling point of water
- 90
- 80
- 70 ← 70 °C = the temperature shown on this thermometer.
- 60
- 50
- 40 ← 37 °C = temperature of a healthy human body
- 30
- 20
- 10
- 0 ← 0 °C = freezing point of water
- −10
- −20 ← −18 °C = temperature of a home freezer

1 Write the temperature shown on each thermometer.

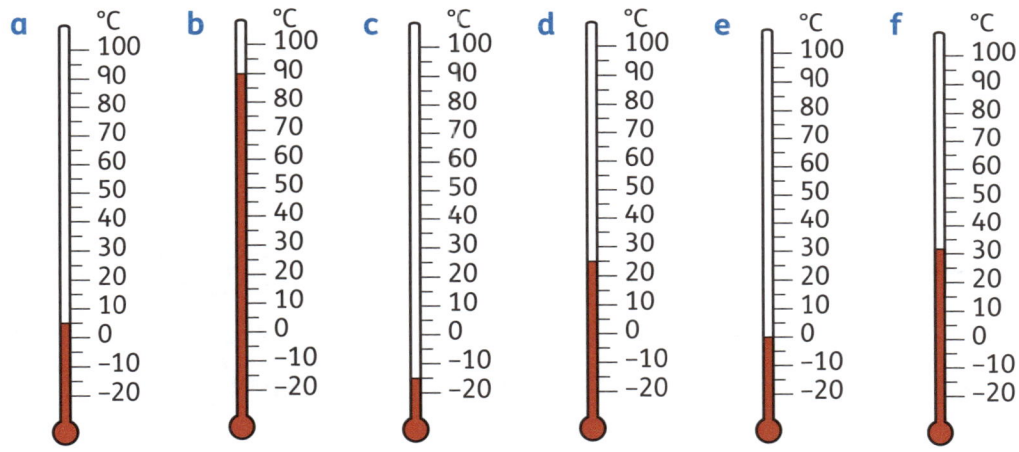

2 Write the temperatures from question 1 in order from coldest to hottest.

Problem solving

Use a number line.

3 Oymyakon and Verkhoyansk are places in Russia. The lowest temperature recorded at Oymyakon was −67.8 °C and at Verkhoyansk it was −69.8 °C.

a Which place recorded the lower temperature?

b What is the difference between these two temperatures?

Temperature changes

When things gets warmer, we say their *temperature rises*. When they get cooler, we say their *temperature falls*. In thermometers that use liquid, you can see the liquid going up the scale (rising) when it gets warmer and down the scale (falling) when it gets colder.

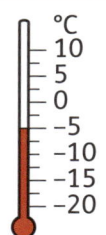

Temperature was
−5 °C

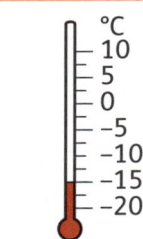

Temperature falls
10 °C to −15 °C

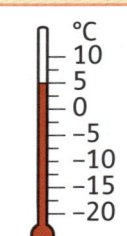

Temperature rises
20 °C to 5 °C

1 Write the temperature that will show on a thermometer after each of these changes:

   a was 13 °C, rises 7 °C   **b** was 8 °C, rises 5 °C

   c was 2 °C, falls 3 °C   **d** was −4 °C, falls 6 °C

   e was −7 °C, falls 5 °C   **f** was −1 °C, rises 5 °C

 Problem solving

> Draw number lines to show the changes and work backwards if you need to.

2 At 7 p.m., the temperature outside was 3 °C.
At midnight it was 5 °C colder and by 4 a.m. the temperature had dropped another 2 °C.
What was the temperature at 4 a.m.?

3 A cold wind makes it feel colder than the actual temperature.
This is called the wind-chill temperature.
On a high mountain, the wind chill temperature was −12 °C.
This was 9 °C colder than the actual temperature.
What was the actual temperature?

4 At 7 p.m. the thermometer showed −6 °C. This was 9 °C colder than it had been at 2 p.m.
The temperature at 2 p.m. was 7 °C warmer than it had been at 7 a.m. What was the temperature at 7 a.m.?

UNIT 14 Perimeter and area

Perimeter

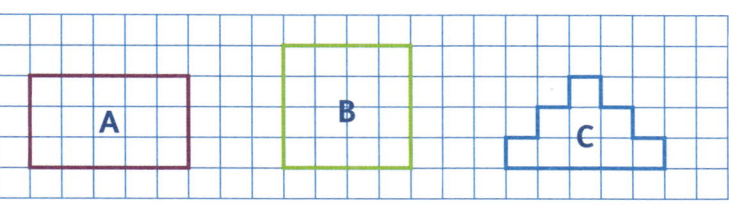

> ### Think and share
>
> Perimeter is the distance around a closed shape.
>
> - Use your finger to trace round the perimeter of each of these shapes.
> - Which one do you think has the greatest perimeter? Why?
>   - What would you measure to find the perimeter of each shape?
>   - Work out the perimeter of each shape. Do the results surprise you?
>   - Marcia says you can find the perimeter of a rectangle without measuring if you know its length and width. How would she be able to do that? Write a rule using L for length and W for width.
>   - Can you make up a rule for finding the perimeter of a square without adding up the lengths of the sides?

1. Naadira cut these shapes out of 1-cm grid paper. Calculate the perimeter of each shape.

   a    b 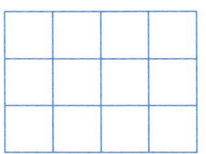   c

   d 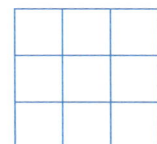   e

2. Draw these rectangles in your book. Write the perimeter of each one next to it.

   a length 6 cm, width 3 cm   b length 9 cm, width 2 cm
   c length 10 cm, width 4 cm   d length 3 cm, width 1 cm

➡ Workbook page 59 and page 60

Calculate perimeter

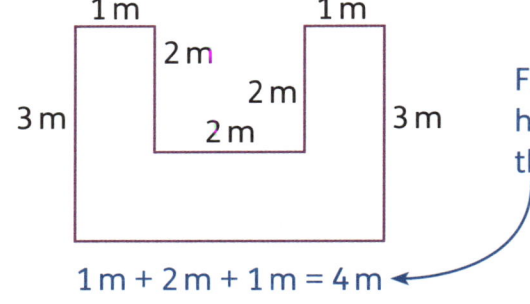

Work out the perimeter of this shape.

1 m   1 m
2 m
2 m
3 m   2 m   3 m

First work out the missing horizontal length across the bottom.

Then calculate the perimeter by adding the lengths of the sides.

1 m + 2 m + 1 m = 4 m

Perimeter = 3 m + 1 m + 2 m + 2 m + 2 m + 1 m + 3 m + 4 m = 18 m

In diagrams like this, you cannot find the perimeter by measuring, because the diagram does not show the real-life size of the shape.

You may need to work out missing measurements using other known measurements before you can calculate the perimeter.

1 Work out the lengths of any sides that are not given. Then calculate the perimeter of each shape.

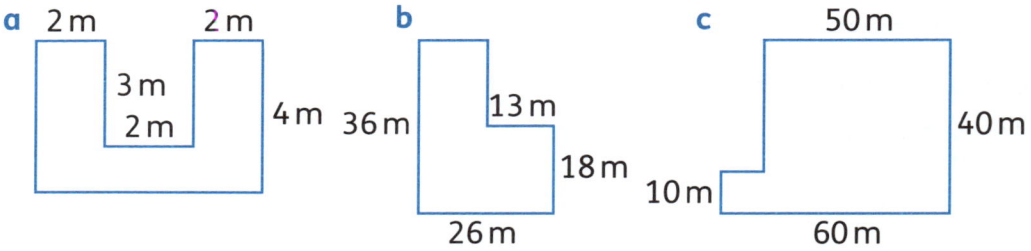

a 2 m   2 m
3 m
2 m

b 4 m   36 m   13 m   18 m   26 m

c 50 m   40 m   10 m   60 m

2 In these shapes, the perimeter is given. Work out the missing lengths of the sides marked with a letter in each shape.

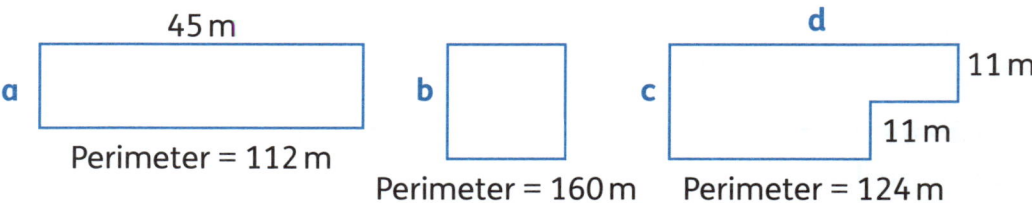

a 45 m
Perimeter = 112 m

b Perimeter = 160 m

c **d** 11 m   11 m
Perimeter = 124 m

➡ *Workbook page 61*

Area

Area is the amount of space inside or covered by a 2D shape.

This shape has been drawn on a grid of squares.

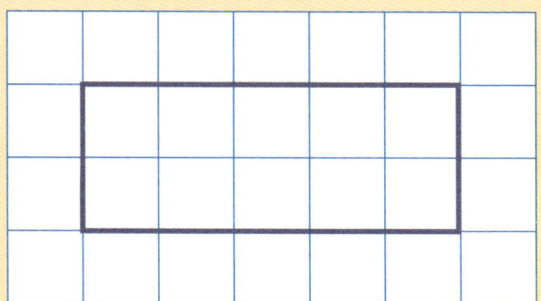

If you count the squares inside the shape you will see that it covers an area of 10 squares.

The squares on the grid all have sides that are 1 cm long.

A square with sides that are 1 cm long is called a **square centimetre**.

We can write this as 1 square centimetre or in a short way as 1 cm^2. The little 2 written above the unit means 'squared'.

The rectangle on the grid has an area of 10 cm^2.

1 These shapes are drawn on a 1-cm grid. Count the squares to find the area of each shape.

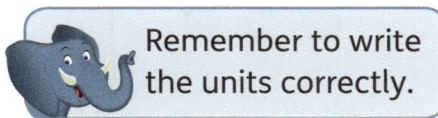

Remember to write the units correctly.

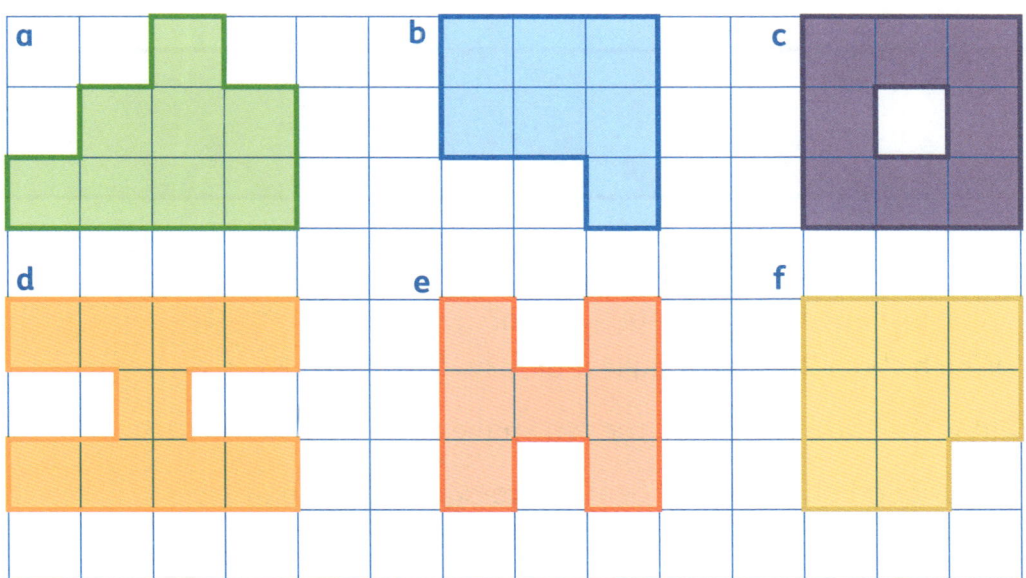

▶ *Workbook page 62*

More area

1 These shapes are drawn on a 1-cm grid. Find the area of each shape in square centimetres.

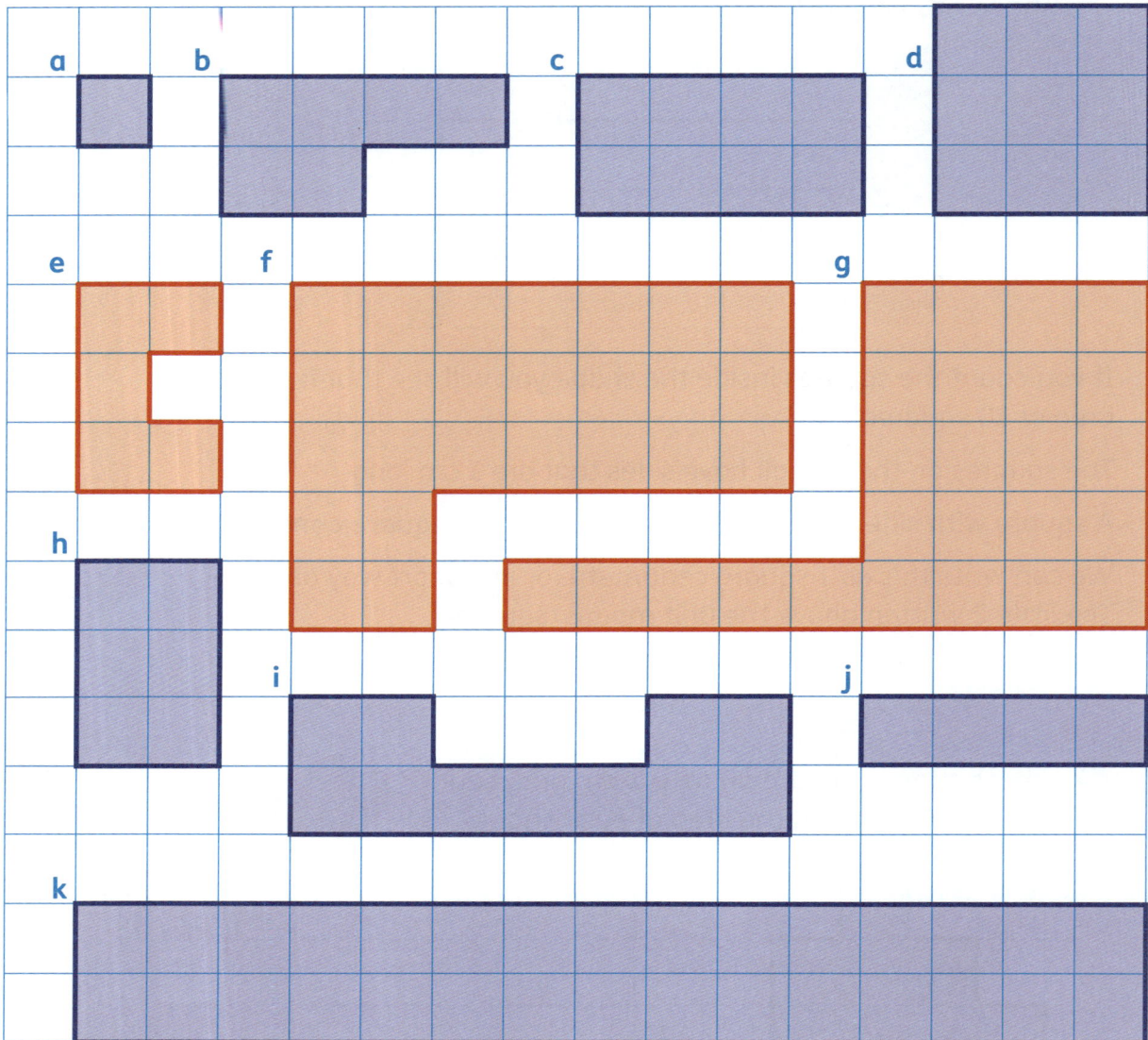

2 Compare the areas of the shapes.
- **a** Which shape has the greatest area?
- **b** Which shape has the smallest area?

3 What is the total area of all the red shapes?

4 Put your hand on a piece of centimetre-squared paper.
Use a pencil to draw around your hand.
- **a** Estimate the area of your hand in square centimetres.
- **b** Draw a triangle with an area equal to the estimated area of your hand.

➡ *Workbook page 63*

Fractions

Revisit fractions

💭 Think and share

What do you remember about fractions? Tell your partner three things.

- Match each diagram to one of the fractions in the box. Explain how you worked it out.

A    B    C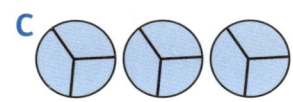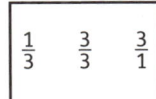

- Mario says the arrow on this number line shows $\frac{1}{7}$ because there are seven small divisions on the number line. What mistake has Mario made?

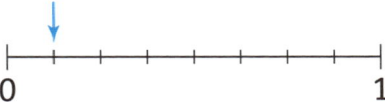

These four containers are all the same size, but they are not full. The fraction of each container that is filled is shown.

 $\frac{2}{3}$ $\frac{7}{8}$ $\frac{6}{12}$ $\frac{2}{5}$

- Which containers are more than half-full?
- Which container is exactly half-full? Why isn't it labelled $\frac{1}{2}$?

1 Draw strips on squared paper and colour them to show these fractions.

a $\frac{1}{2}$ b $\frac{3}{4}$ c $\frac{2}{3}$ d $\frac{5}{8}$

e $\frac{4}{5}$ f $\frac{7}{8}$ g $\frac{3}{10}$ h $\frac{8}{10}$

2 Use your coloured strips to help you find the missing fractions in these calculations.

a $\frac{1}{2} + \boxed{} = 1$ b $\frac{3}{4} + \boxed{} = 1$ c $\frac{2}{3} + \boxed{} = 1$

d $\frac{5}{8} + \boxed{} = 1$ e $\frac{4}{5} + \boxed{} = 1$ f $\frac{7}{8} + \boxed{} = 1$

g $\frac{7}{10} + \boxed{} = 1$ h $\frac{2}{10} + \boxed{} = 1$ i $\frac{1}{3} + \boxed{} = 1$

➡️ *Workbook page 64 and page 65*

Compare and order fractions

Look at the shaded part of each shape.

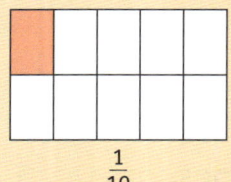 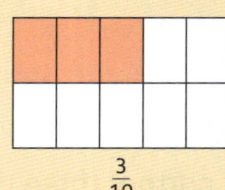 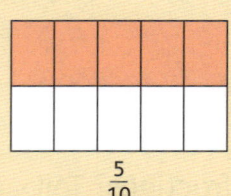 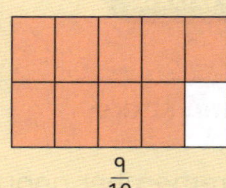

$\frac{1}{10}$ $\frac{3}{10}$ $\frac{5}{10}$ $\frac{9}{10}$

$\frac{1}{10}$ is less than $\frac{3}{10}$ You can write $\frac{1}{10} < \frac{3}{10}$

$\frac{9}{10}$ is greater than $\frac{3}{10}$ You can write $\frac{9}{10} > \frac{3}{10}$

1 Copy the number sentences. Fill in < or > to compare the fractions.

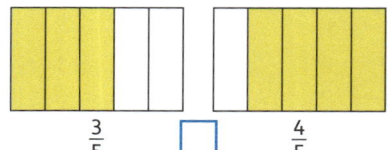

$\frac{3}{5}$ ☐ $\frac{4}{5}$

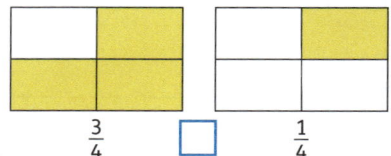

$\frac{3}{4}$ ☐ $\frac{1}{4}$

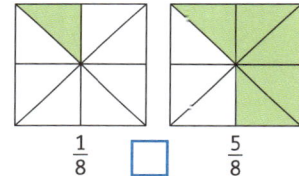

$\frac{1}{8}$ ☐ $\frac{5}{8}$

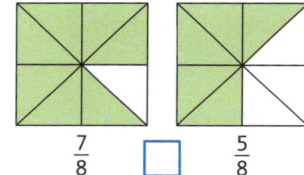

$\frac{7}{8}$ ☐ $\frac{5}{8}$

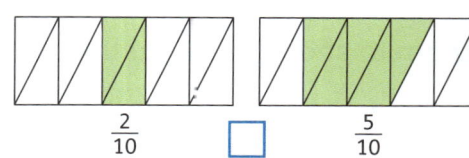

$\frac{2}{10}$ ☐ $\frac{5}{10}$

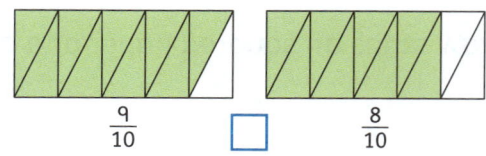

$\frac{9}{10}$ ☐ $\frac{8}{10}$

2 **a** Write the fraction of each shape that is shaded in each set.

   b Write each set of fractions in order, from greatest to smallest.

A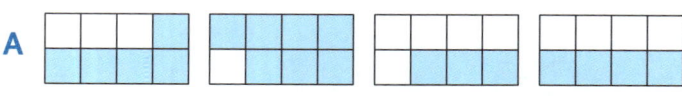

B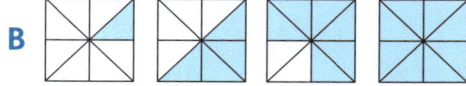

C

Equivalent fractions

Equivalent fractions have the same value.

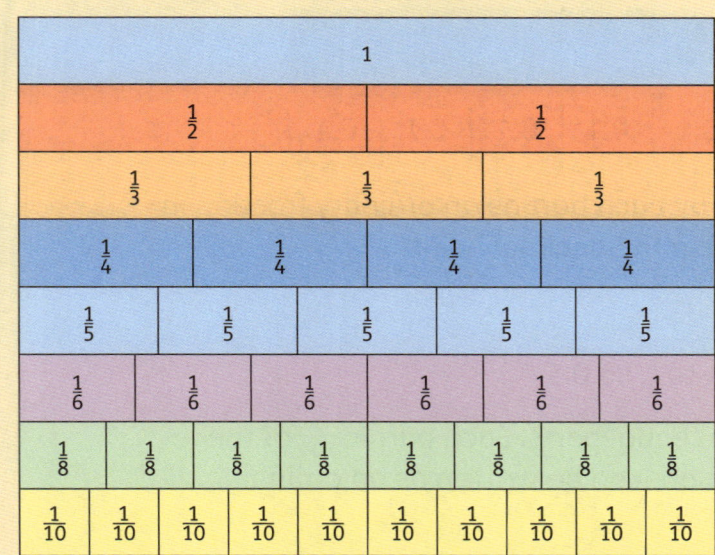

$\frac{1}{2} = \frac{2}{4} = \frac{4}{8}$

$\frac{1}{4} = \frac{2}{8}$

$\frac{3}{4} = \frac{6}{8}$

1 What do these two sets of number lines show? Discuss this in groups.

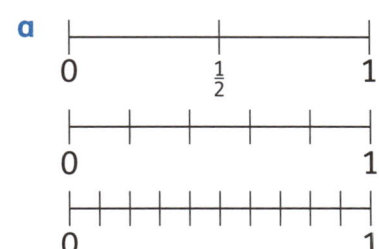

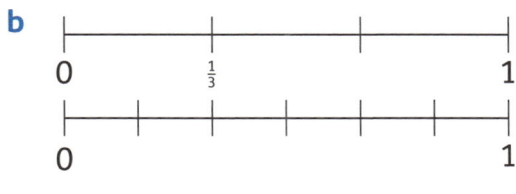

2 Find and write an equivalent fraction for each fraction given.

Use the fraction wall or the number lines.

a $\frac{1}{6}$ **b** $\frac{3}{6}$ **c** $\frac{2}{3}$ **d** $\frac{4}{6}$

e $\frac{8}{10}$ **f** $\frac{2}{8}$ **g** $\frac{2}{5}$ **h** $\frac{6}{8}$

i $\frac{1}{5}$ **j** $\frac{6}{10}$ **k** $\frac{3}{6}$ **l** $\frac{3}{5}$

3 Copy the statements. Fill in <, > or = to make each statement true.

a $\frac{1}{2} \square \frac{3}{4}$ **b** $\frac{3}{4} \square \frac{6}{8}$ **c** $\frac{5}{10} \square \frac{2}{5}$ **d** $1 \square \frac{9}{10}$

4 Rewrite the fractions:

a $\frac{2}{8}, \frac{6}{8}, \frac{1}{2}$ as quarters **b** $\frac{8}{10}, \frac{6}{10}, \frac{2}{10}$ as fifths

➡ *Workbook page 66 and page 67*

Fractions and equivalent decimals

These strips are both divided into tenths.

$\frac{1}{10}$	$\frac{1}{10}$	$\frac{1}{10}$	$\frac{1}{10}$	$\frac{1}{10}$	$\frac{1}{10}$	$\frac{1}{10}$	$\frac{1}{10}$	$\frac{1}{10}$	$\frac{1}{10}$
0.1	0.1	0.1	0.1	0.1	0.1	0.1	0.1	0.1	0.1

We can write the fractions shown by each part as an ordinary fraction in tenths or as a decimal fraction using the decimal point.

$\frac{3}{10} = 0.3$

$\frac{5}{10} = 0.5$ but $\frac{5}{10}$ is also equivalent to $\frac{1}{2}$, so $0.5 = \frac{1}{2}$

These squares are divided into 100 equal parts. Each part is $\frac{1}{100}$ of the square. The shaded part of the square can be written as an ordinary fraction and as a decimal.

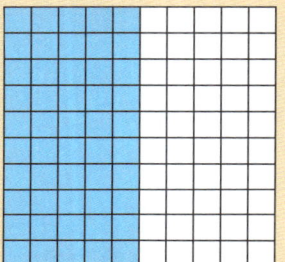
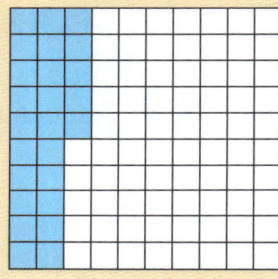
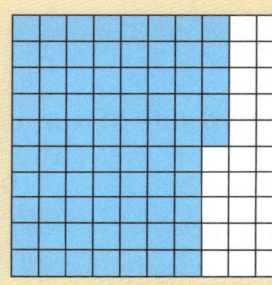

$\frac{50}{100} = 0.5$ $\frac{25}{100} = 0.25$ $\frac{75}{100} = 0.75$

$\frac{50}{100} = \frac{1}{2}$ $\frac{25}{100} = \frac{1}{4}$ $\frac{75}{100} = \frac{3}{4}$

$\frac{1}{2} = 0.5$ $\frac{1}{4} = 0.25$ $\frac{3}{4} = 0.75$

1 Write a decimal that is equivalent to each fraction.

     a $\frac{1}{10}$ **b** $\frac{7}{10}$ **c** $\frac{75}{100}$

     d $\frac{3}{10}$ **e** $\frac{5}{10}$ **f** $\frac{25}{100}$

2 In each part c – c, match the fractions with their decimal equivalents. Write your answers using equals signs like this: $\frac{1}{2} = 0.5$

     a $\frac{8}{10}$ 0.5   **b** $\frac{1}{2}$ 0.25   **c** $\frac{1}{2}$ 0.8

        $\frac{3}{10}$ 0.8 $\frac{3}{4}$ 0.5 $\frac{9}{10}$ 0.5

        $\frac{5}{10}$ 0.3 $\frac{25}{100}$ 0.75 $\frac{80}{100}$ 0.9

➡ *Workbook page 68*

Mixed numbers

How many whole sandwiches can you make with these halves?

There are 5 halves.

You can make $2\frac{1}{2}$ sandwiches.

$2\frac{1}{2}$ is a **mixed number**. It contains both a whole number and a fraction.

We can show $2\frac{1}{2}$ on a number line like this:

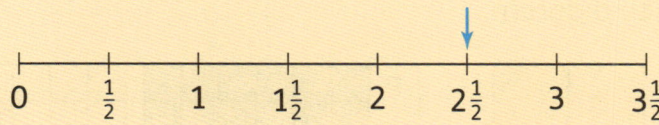

$$0 \quad \frac{1}{2} \quad 1 \quad 1\frac{1}{2} \quad 2 \quad 2\frac{1}{2} \quad 3 \quad 3\frac{1}{2}$$

1 What fraction is coloured? Write the answers as mixed numbers.

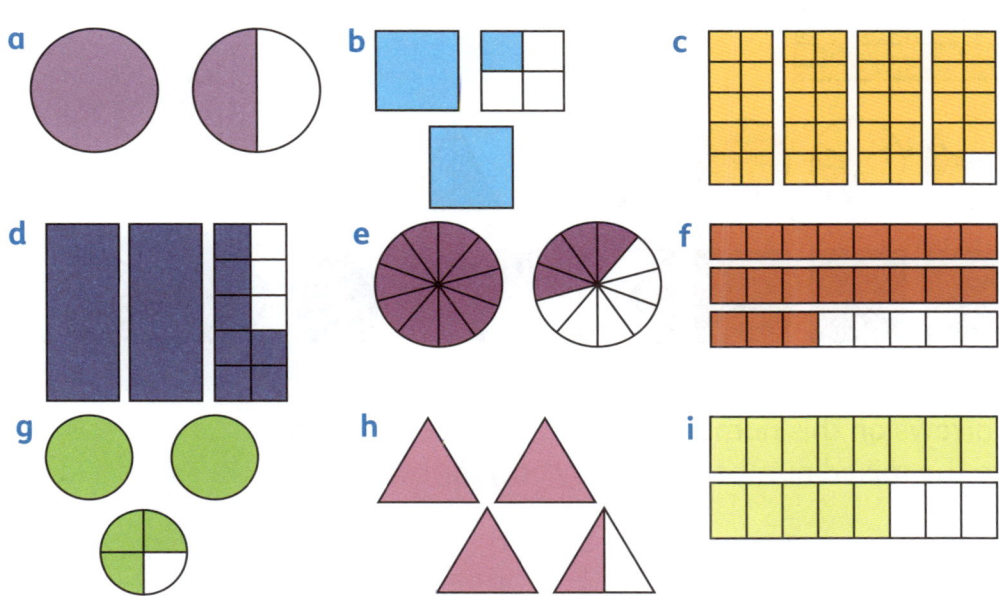

2 Draw number lines to show the position of each mixed number.

a $2\frac{1}{2}$ b $5\frac{3}{4}$ c $1\frac{2}{3}$

d $3\frac{1}{8}$ e $1\frac{1}{10}$ f $2\frac{3}{5}$

➡ *Workbook page 69*

Mixed numbers and improper fractions

Each pie is cut into thirds.

If you look at the pieces of pie you can see that there are eight thirds.

We can write this as $\frac{8}{3}$ (we say eight thirds).

$\frac{8}{3}$ is more than $\frac{3}{3}$ (it is greater than 1, because $1 = \frac{3}{3}$) so we say it is an **improper fraction**.

$2\frac{2}{3} = \frac{8}{3}$

1 Write an improper fraction and a mixed number to describe the parts shaded in each diagram.

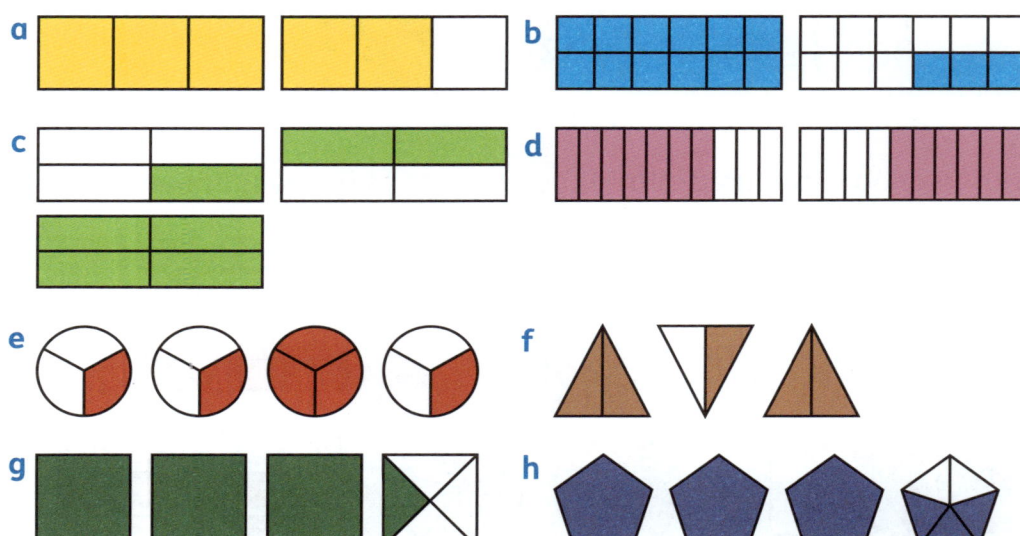

2 The arrows on this number line indicate mixed numbers. Write the mixed number shown by each arrow.

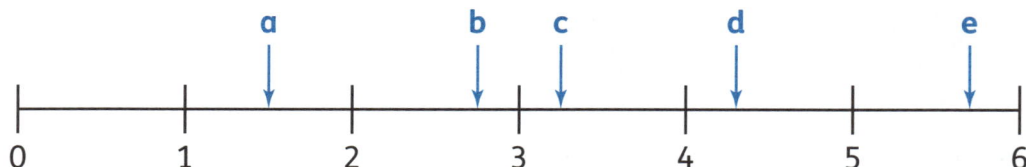

Write improper fractions as mixed numbers

It is often easier to work with improper fractions if you write them as mixed numbers.

These examples show two ways of writing $\frac{7}{6}$ as an equivalent mixed number.

Using diagrams

$\frac{7}{6} =$

This is $\frac{6}{6} + \frac{1}{6}$

$= 1 + \frac{1}{6}$

$= 1\frac{1}{6}$

Using division

$\frac{7}{6}$ means $7 \div 6$

7 divided by 6 is 1 with 1 left over.

The 1 that is left over is 1 of 6 pieces or $\frac{1}{6}$

$\frac{7}{6} = 1\frac{1}{6}$

1 Write each of these improper fractions as a mixed number.

a $\frac{3}{2}$　　　　b $\frac{4}{3}$　　　　c $\frac{5}{4}$　　　　d $\frac{7}{4}$

e $\frac{9}{4}$　　　　f $\frac{15}{7}$　　　g $\frac{12}{5}$　　　h $\frac{13}{3}$

i $\frac{19}{4}$　　　j $\frac{7}{3}$　　　　k $\frac{11}{6}$　　　l $\frac{25}{6}$

💡 **Problem solving**

2 Which fraction is greater in each of these pairs?

a $5\frac{1}{2}$ or $\frac{7}{2}$　　　b $2\frac{3}{4}$ or $\frac{7}{4}$　　　c $1\frac{3}{5}$ or $\frac{9}{5}$

d $3\frac{2}{7}$ or $\frac{17}{7}$　　e $\frac{4}{3}$ or $1\frac{2}{3}$　　f $\frac{14}{4}$ or $3\frac{3}{4}$

g $7\frac{1}{2}$ or $\frac{20}{2}$　　h $\frac{18}{8}$ or $2\frac{3}{8}$　　i $12\frac{1}{3}$ or $\frac{40}{3}$

j $\frac{19}{5}$ or $3\frac{3}{5}$　　k $7\frac{2}{7}$ or $\frac{48}{7}$　　l $\frac{50}{8}$ or $6\frac{7}{8}$

> Change the fractions so they are both mixed numbers or both improper fractions to compare them.

3 Write each set of fractions in order from smallest to greatest.

a $1\frac{1}{2}$　　　$\frac{7}{3}$　　　$\frac{9}{5}$　　　$2\frac{1}{4}$　　　$\frac{15}{6}$

b $\frac{9}{2}$　　　$\frac{3}{4}$　　　$\frac{12}{5}$　　　$\frac{19}{6}$　　　$\frac{21}{5}$

c $\frac{12}{5}$　　　$\frac{7}{3}$　　　$\frac{4}{3}$　　　$\frac{9}{4}$　　　$\frac{12}{8}$

▶ *Workbook page 70*

Add and subtract fractions

You can add and subtract fractions and whole numbers. Your answer could be a mixed number or an improper fraction.

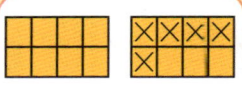

$2 - \frac{5}{8} = 1\frac{3}{8}$

$\frac{7}{6} + \frac{9}{6} = \frac{16}{6}$ or $2\frac{4}{6}$

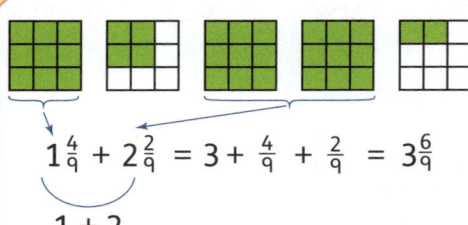

$1\frac{4}{9} + 2\frac{2}{9} = 3 + \frac{4}{9} + \frac{2}{9} = 3\frac{6}{9}$

$1 + 2$

1 Calculate. Draw diagrams if you need to.

a $\frac{2}{7} + \frac{4}{7}$

b $\frac{5}{9} + \frac{4}{9}$

c $\frac{2}{5} + \frac{4}{5}$

d $\frac{11}{12} + \frac{4}{12}$

e $4\frac{1}{3} + 2\frac{1}{3}$

f $\frac{4}{7} + \frac{3}{7} + \frac{2}{7}$

g $\frac{11}{9} + \frac{4}{9}$

h $\frac{2}{3} + \frac{4}{3}$

i $\frac{8}{9} - \frac{3}{9}$

j $\frac{11}{12} - \frac{5}{12}$

k $3 - \frac{1}{3}$

l $\frac{7}{4} - \frac{1}{4}$

m $3\frac{2}{3} - \frac{1}{3}$

n $\frac{9}{10} - \frac{2}{10}$

o $3 - \frac{5}{7}$

p $\frac{5}{7} - \frac{4}{7}$

2 Four pupils added fractions and got these answers.
Write two possible sums for each answer.

a $\frac{9}{10}$

b $2\frac{1}{3}$

c $\frac{11}{12}$

d $\frac{19}{5}$

3 What was subtracted in each calculation?

a $3 - \square = 2\frac{1}{2}$

b $\frac{15}{12} - \square = \frac{7}{12}$

c $\frac{13}{10} - \square = 1$

💡 Problem solving

4 A cake is cut into 12 equal slices. Jess takes $\frac{1}{4}$, Misha takes 2 slices and Nia takes 4 slices. What fraction of the cake is left?

5 Johan used 2 whole bags and $\frac{5}{6}$ of another bag of bird seed during the week. Over the weekend he used another $\frac{5}{6}$ of a bag. How much bird seed did he use altogether?

6 Malik cycled $12\frac{3}{5}$ km. Sajid cycled $11\frac{4}{5}$ km. How much further did Malik cycle?

➡ *Workbook page 71*

Calculate with fractions

Work out $\frac{2}{5}$ of £35.

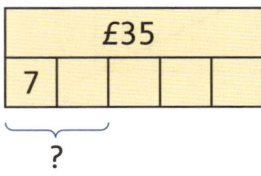

You can find $\frac{1}{5}$ by dividing. $35 \div 5 = 7$

$\frac{1}{5} = 7$, so $\frac{2}{5} = 2 \times 7 = 14$

$\frac{2}{5}$ of £35 = £14

I earned some money selling crafts at the market. It cost £8 to rent the stall. This is $\frac{1}{6}$ of the money I made. How much money did I make?

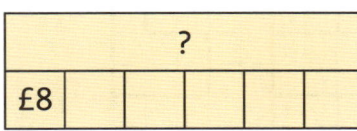

$\frac{1}{6}$ of the money = £8.

The whole amount is $\frac{6}{6}$.

$\frac{6}{6} = 6 \times 8 = £48$

I made £48.

Sondra has used 12 beads from a full packet. This is $\frac{2}{9}$ of the beads in the packet. How many beads are left?

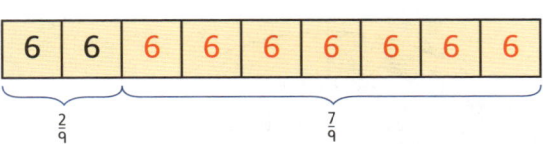

$\frac{9}{9} - \frac{2}{9} = \frac{7}{9}$ beads are left.

$\frac{2}{9} = 12$, so $\frac{1}{9} = 6$

$\frac{1}{9} = 6$, so $\frac{7}{9} = 7 \times 6 = 42$

There are 42 beads left.

💡 Problem solving

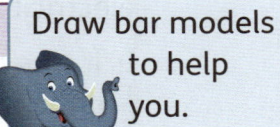

Draw bar models to help you.

1. Maria has 24 markers and crayons. $\frac{2}{3}$ of them are crayons. How many markers does she have?

2. Priya spent $\frac{2}{5}$ of her pocket money on books. She had £9 left. How much did the books cost?

3. Saleem has $\frac{5}{11}$ of a packet of sweets left. When he counts them he gets 20. How many were in the packet to start with?

4. Write a problem to match each model and show how you would solve it.

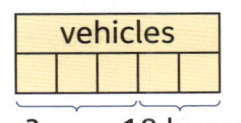

? cars 18 buses

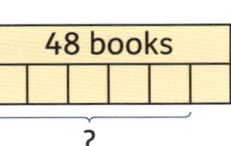

?

Mixed practice 2

1 **a** Explain why the red line on this rectangle is *not* a line of symmetry.

b Draw any rectangle in your book. Draw a correct line of symmetry on it.

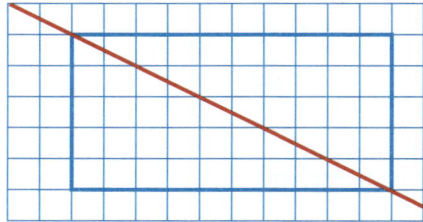

2 The red line on each diagram is a mirror line.

a Which patterns are symmetrical? Write the letters.

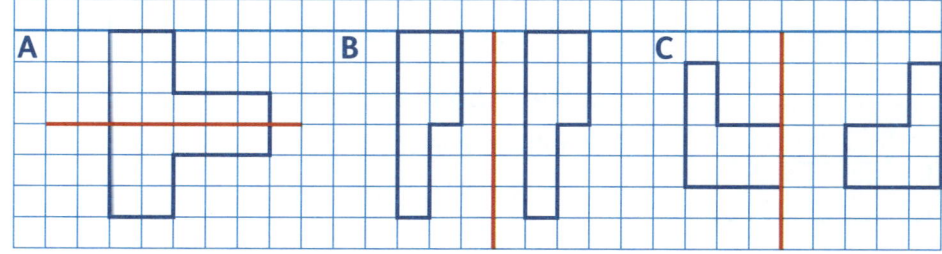

b Which shape has an area of exactly 8 squares?

3 Mike drew this chart to show how electricity is used in his home.

a What fraction of the electricity is used by the TV and computer?

b Which item uses $\frac{1}{3}$ of the electricity?

c Is it true that the lights use $\frac{1}{6}$ of the total electricity used?

d This home used 240 units of electricity. How many units were used for heating and cooking?

Electricity use in my home

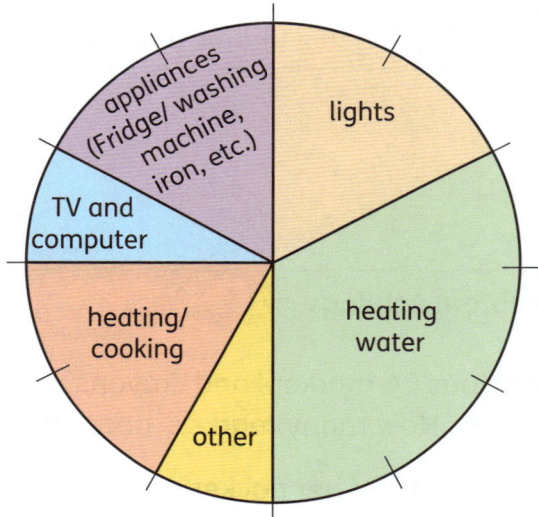

e In one month, the total electricity used by the appliances, TV and computer, and heating and cooking, was 105 units. How many units were used in the home altogether?

4 Look at the diagram.

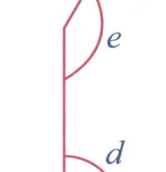

a What type of polygon is this?

b Milla says the sides are all the same length, so this is a regular polygon. Is she correct? Explain your answer.

c Which of the marked angles are obtuse?

d What type of angle is a?

e How many right angles are in this shape? Write the letters of the angles which are right angles.

f Each side is $2\frac{1}{2}$ cm long. What is the perimeter of the shape?

5 Estimate and then calculate each answer. Use an inverse operation to check that each calculation is correct.

a 287 – 129

b 632 + 910

c 4123 + 289

d 876 – 499

6 Three pupils recorded the different types of vehicles that passed their school during break time. They made this tally table of their results.

a How many more buses passed the school than trucks?

b How many vehicles passed the school altogether?

c The pupils drew bar charts to show the data. Write down two things that are wrong with each chart.

Type of vehicle	Tally
Bicycle	卌
Truck	卌 卌 卌 |||
Bus	卌 卌 卌 卌 ||
Car	卌 卌 卌 卌 卌 卌 卌

A

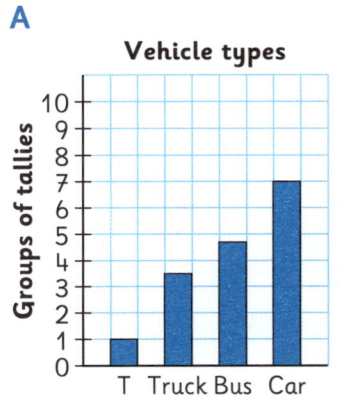

B

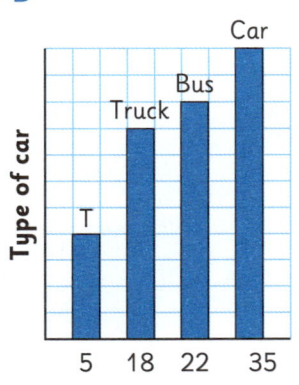

C

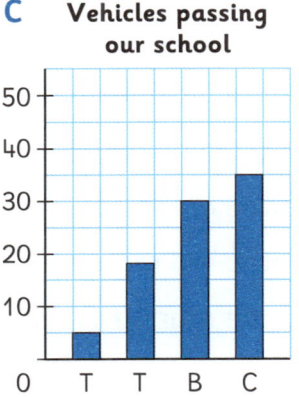

Position and movement

Describe positions on a grid

Think and share

How can you describe the position of **points** A, B, C and D on this **coordinate grid**?

Read these statements. Find the things they describe on the coordinate grid.

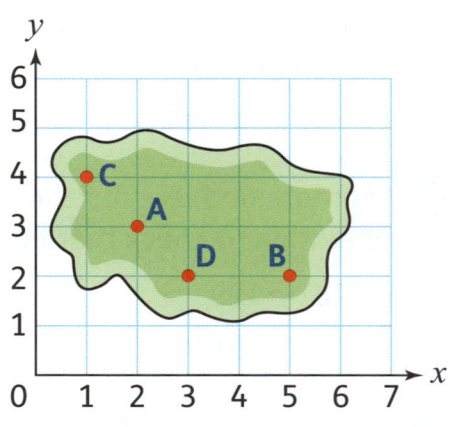

- The grid is formed by two number lines called axes. There is an *x*-**axis** and a *y*-**axis**.

- The axes both start from 0. The point where they start is (0, 0).

- Each position on the grid can be described using two numbers, for example, point B is at (5, 2).

Describe the positions of points A, C and D.

1 What shape is found at each of these positions?

    a (2, 2)   **b** (6, 1)   **c** (1, 6)

    d (4, 6)   **e** (6, 4)   **f** (5, 4)

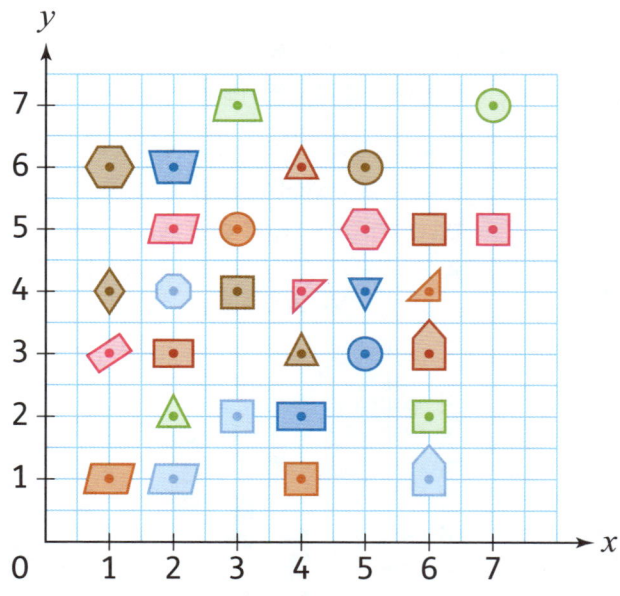

2 Give the position of:

    a the blue circle

    b the pink hexagon

    c the pink parallelogram

    d the green square

    e the red rectangle

    f the orange circle.

3 There is only one shape with eight sides on the grid.

    Give its name, colour and position.

Use coordinates

Each position can be given using two **coordinates**.

- The coordinates are written between brackets as a pair. For example, point A is in position (1, 4).

- The coordinates indicate where a horizontal and a vertical line cross each other.

- The order of the points matters. Point A is at (1, 4), Point C is at (4, 1).

- The horizontal value (x-coordinate) is always given first.

- The vertical value (y-coordinate) is given second.

Can you give the positions of the other points on the grid?

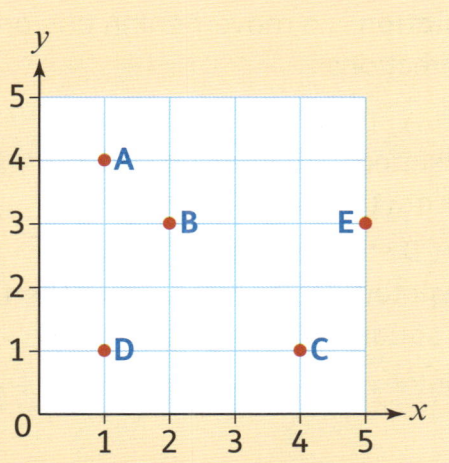

Remember, x comes before y in the alphabet.

1 What point is found in each of these positions? Write the letters only.

a (9, 1)   b (1, 7)

c (6, 5)   d (6, 3)

e (3, 6)   f (7, 8)

2 Give the coordinates of these points.

a B   b C

c E   d G

e I   f N

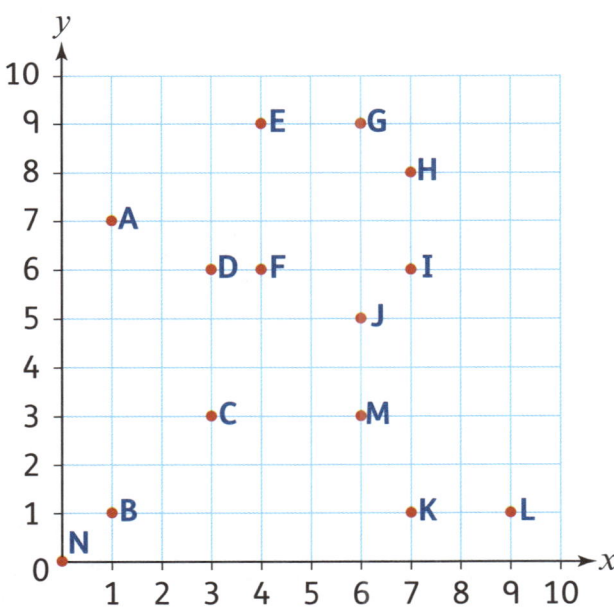

➡ Workbook page 72

Translations

A **translation** is a movement in one or more directions.

- Point A started at (1, 1). It was translated 3 squares up and is now at (1, 4).
- Point D started at (4, 1). It was translated one square up and then one square to the right. It is now at (5, 2).

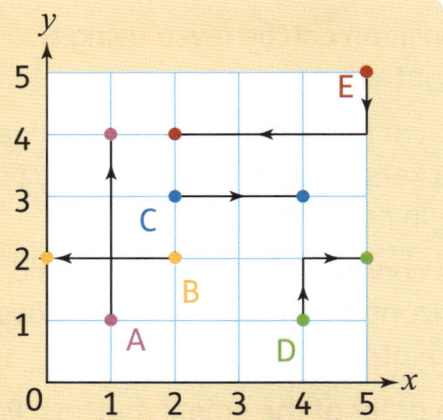

Use the grid below to answer questions 1 to 5.

1 Translate Points A, B and C each 2 squares right and 1 square down. Write the coordinates of each point after the translation.

2 Describe the translation that would move Point C to Point D.

3 How could you translate Point E to Point B?

4 Translate Point H 2 squares down, 1 square left and then 2 squares down. What are Point H's new coordinates?

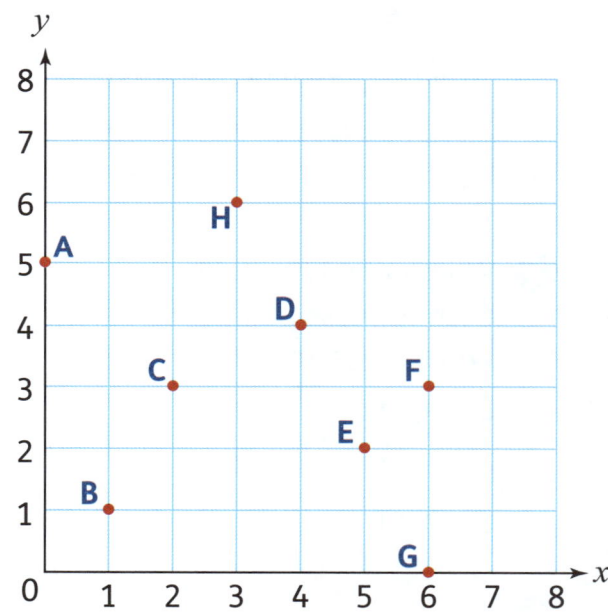

Use the grid above and work backwards.

💡 **Problem solving**

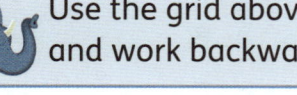

5 Points F and G are each translated 2 squares up or down and 2 squares left or right. What could the new coordinates of each point be? Give two possible answers for each.

6 After a translation of 3 squares right and 4 squares down, a point is at (8, 3). What are the coordinates of its starting position?

➡ *Workbook page 73*

Shapes on a grid

You can plot points on a grid and join them to make different shapes.

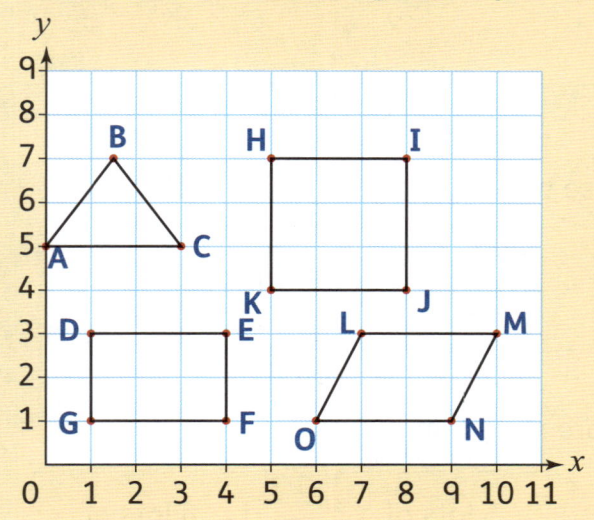

1. Mishak plots the points (2, 2), (4, 0), (4, 4) and (6, 2) and joins them in order. What shape does he make?

2. Marija says that the points (0, 9), (6, 9), (6, 6) and (2, 6) are the vertices of a rectangle. Is she correct?

3. Salman has plotted the points (4, 5), (4, 2) and (9, 2). What type of triangle will he get if he joins the points? How do you know this?

4. Look at the shapes on the grid above. Translate each shape 1 square down and 2 squares right. Write the coordinates of the vertices of each translated shape.

Problem solving

5. One side of a square has been drawn on this grid. One of the other vertices of the square is at (1, 2).

   a What are the coordinates of the fourth vertex?

   b The whole square is translated 1 square left and 1 square up. What are the new coordinates of the vertices?

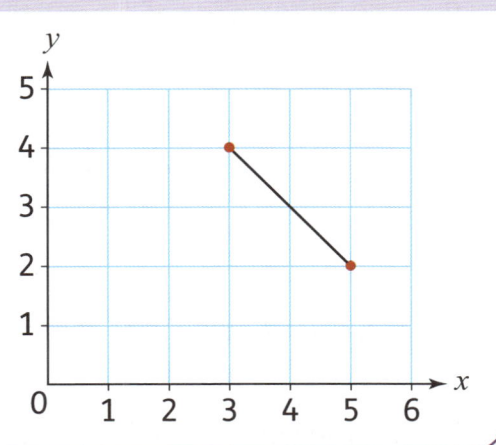

Workbook page 74 and page 75

Multiplication

Remember your facts

Think and share

Nisha uses sets of matching cards like these to revise multiplication facts.

array	in words	facts	product
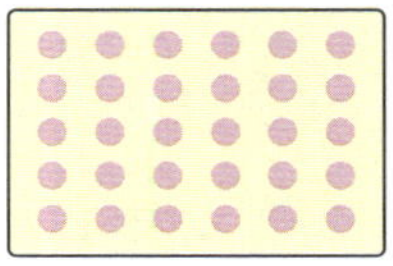 (array of dots)	5 lots of 6	5 × 6 / 6 × 5	30

There are five sets of cards here but they are mixed up.

a 42

b 3 lots of 8

c 5 lots of 8

d 6 lots of 7

e 3 × 8 / 8 × 3

f (array of green dots)

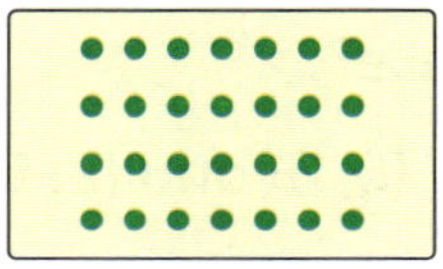

g 40

h 4 lots of 7

i 4 × 9 / 9 × 4

j (array of dots)

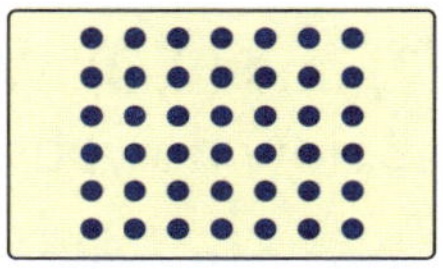

k 24

l 4 × 7 / 7 × 4

m 28

n 6 × 7 / 7 × 6

o 4 lots of 9

p (array of red dots)

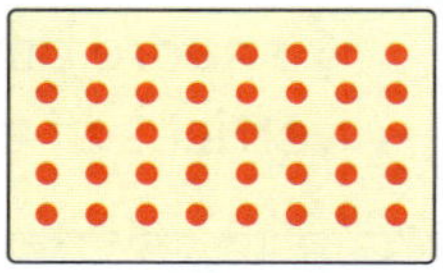

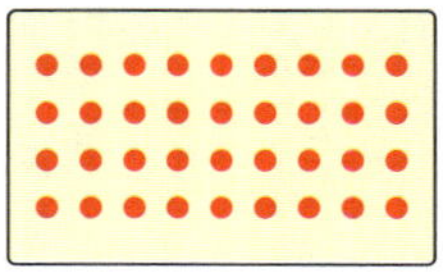

q (array of red dots)

r 36

s (array of dots)

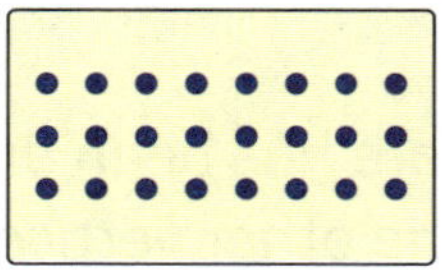

t 5 × 8 / 8 × 5

Decide which cards go together.

Keep a record of your work.

1 Make your own cards for ten different facts from the 6, 7, 9, 11 and 12 times tables.

2 Shuffle your cards. Exchange cards with a partner.
Sort each other's cards into matching fact sets.

➡ *Workbook page 76*

Multiply tens

What is 50×3?

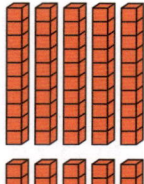

50 is five lots of 10

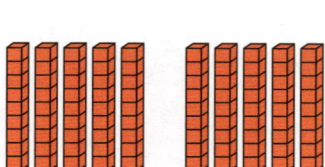

3×5 lots of 10 is 15 lots of 10

So, 3×50 is the same as $3 \times 5 \times 10$

$3 \times 5 \times 10 = 15 \times 10 = 15$ tens or 150

You could rewrite 50×3 as 3×50 to make the multiplication easier.

How can knowing multiplication facts help you multiply by any number of tens?

1 Write these amounts in numerals.

   a 7 tens   b 9 tens   c 11 tens

   d 21 tens   e 28 tens   f 30 tens

   g 50 tens   h 72 tens   i 99 tens

2 What is:

   a 6×3 tens   b 3×9 tens   c 4×5 tens

   d 5×6 tens   e 3×3 tens   f 9×10 tens

3 Calculate.

   a 6×20 b 4×30 c 5×60

   d 40×8 e 50×7 f 60×3

   g 80×9 h 50×5 i 90×4

 Problem solving

Break up numbers to help you multiply.

4 80 planes land at an airport each day.
How many planes land there in a week?

5 A school bus can carry 58 pupils and 2 teachers. How many pupils and teachers can be carried in 9 buses?

Multiply larger numbers by 10

You can use the multiplication facts and the patterns of multiples that you already know to find quick methods of multiplying any number by 10.

$8 \times 10 = 8$ tens $= 80$

$18 \times 10 = 18$ tens $= 180$

$123 \times 10 = 123$ tens $= 1230$

> Remember all whole number multiples of 10 have a 0 in the ones place.

We can show these numbers on place-value tables:

Thousands	Hundreds	Tens	Ones
			8
		8	0

Thousands	Hundreds	Tens	Ones
		1	8
	1	8	0

Thousands	Hundreds	Tens	Ones
	1	2	3
1	2	3	0

When you multiply a whole number by 10, each digit moves one place to the left on the place-value table and you write a 0 as a place holder in the ones place.

1 Write these amounts in numerals.

a 45 tens   b 67 tens   c 91 tens

d 146 tens   e 234 tens   f 420 tens

2 Try to do these multiplications mentally.

a 148×10 b 185×10 c 177×10

d 219×10 e 209×10 f 290×10

g 306×10 h 360×10 i 366×10

➡ *Workbook page 77*

Multiply by 100

7 hundreds = 700, so

$7 \times 100 = 700$

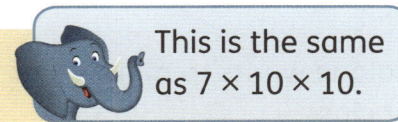 This is the same as $7 \times 10 \times 10$.

| 100 tissues | 100 tissues | 100 tissues | 100 tissues | 100 tissues | 100 tissues | 100 tissues |

12 hundreds = 1200, so

$12 \times 100 = 1200$

You already know that multiples of 100 have two zeros at the end.

You can use this fact to quickly multiply any numbers by 100.

1 An airline is offering a special deal on flights to celebrate their 100th year in business. The first 100 people who buy tickets online can book a flight to anywhere they want for just £100. Work out the cost of:

  a 2 tickets
  b tickets for a family of 6
  c 9 tickets
  d 11 tickets

2 Write these amounts using numbers.

  a 45 hundreds
  b 88 hundreds
  c 90 hundreds
  d 24 hundreds
  e 15 hundreds
  f 20 hundreds

3 Find the product of these numbers.

  a 3×100
  b 9×100
  c 12×100
  d 13×100
  e 25×100
  f 77×100

Problem solving

4 Tablet computers are on sale for £89. A school buys 99 of these. How much do they cost?

 How can you multiply by 100 to find the answer?

Workbook page 78

Break up numbers and add to multiply

Mrs Malone wants to make 15 of these 9-pin geoboards for her class.

How many nails will she need?

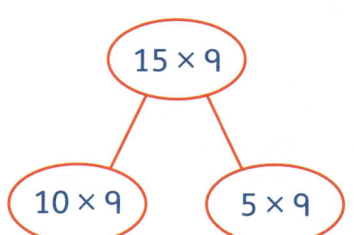

She needs 9 pins for each board, multiplied by 15 boards.

You know that 15 = 10 + 5.

 You can work out 10 × 9 and 5 × 9 using facts you know.

Then you can add the products to find 15 × 9.

Find the product 15 × 9.

$15 \times 9 = (10 \times 9) + (5 \times 9)$

$\quad = \quad 90 \quad + \quad 45$

$\quad = \quad 90 \quad + \quad 10 + 35$

$\quad = \quad 135$

```
        15 × 9
       /      \
   10 × 9    5 × 9
```

1 Find the product.

a 9 × 19	b 9 × 43	c 5 × 82
d 12 × 6	e 29 × 4	f 41 × 8
g 4 × 23	h 51 × 6	i 25 × 8

2 Calculate:

a 4 × 29	b 6 × 38	c 5 × 27
d 31 × 2	e 44 × 6	f 82 × 3
g 91 × 6	h 41 × 5	i 83 × 5

3 Lee works out 63 × 4 like this:

63 + 63 = 126

126 + 126 = 240 + 12 = 252.

a What did Lee do?

b When will this method be useful? Why?

➡ *Workbook page 79*

Use written methods to multiply

You already know some strategies for multiplying whole numbers.

Read through these examples to see different written methods for multiplying 43 × 7.

Always estimate first. Use rounding to do this.

43 rounds to 40.

$7 \times 40 = 7 \times 4 \times 10 = 28 \times 10 = 280$

Partitioning	Grid method	Column method

Partitioning

43 × 7
= (40 × 7) + (3 × 7)
= 280 + 21
= 301

Grid method

×	7
40	280
3	21

280 + 21 = 301

Column method

43
× 7
——
21   3 × 7 = 21
280   40 × 7 = 280
——
301

1 Use the method you find easiest to do these multiplications. Remember to estimate first.

a 81 × 5
b 6 × 28
c 7 × 38
d 7 × 67
e 9 × 88
f 75 × 9
g 8 × 59
h 6 × 49
i 3 × 93

 Problem solving

> Can you use a bar model to help you understand the problem better?

2 Solomon delivers 83 parcels every day of the week. How many does he deliver altogether in a week?

3 Sam delivered 5 boxes (each with 48 bottles of juice) and 3 bags (each with 65 oranges) to a school tuck shop. How many items is this altogether?

4 During the holidays, Li cycled 24 km a day for 7 days. Micah cycled 22 km a day for 8 days.

a Who cycled the furthest?

b How much further did that person cycle?

Use written methods for 3-digit numbers

You can use the methods you already know to multiply even bigger numbers.

Estimate first:
$$400 \times 3 = 4 \times 100 \times 3$$
$$= 1200$$

What is 438×3?

Partitioning

$438 \times 3 = (400 \times 3) + (30 \times 3) + (8 \times 3)$
$= 1200 + 90 + 24$
$= 1290 + 24$ ⟵ It is easier to add $1290 + 24$ in parts.
$= 1314$ $1300 + 14$

Grid method

×	3
400	1200
30	90
8	24

$1200 + 90 + 24 = 1314$

Column method

```
    438
  ×   3
  ─────
     24     8 × 3
     90     30 × 3
   1200     400 × 3
  ─────
   1314
```

1 Use the method you find easiest to do these multiplications. Remember to estimate first.

    a 48×3 **b** 79×9 **c** 142×5 **d** 358×3

    e 297×3 **f** 123×7 **g** 132×2 **h** 463×5

2 Estimate and then solve these problems using a written method.

    a A chicken farm produces 185 eggs each day.

      How many eggs does it produce in a week?

    b Jonas gave three charities £462 each as a donation.

      How much did he donate altogether?

    c Each carriage on a train has 54 seats. How many seats are there if the train has eight carriages?

➡ *Workbook page 80*

Multiply and solve

Read each problem carefully and think about how you can solve it.
Estimate by rounding before you start so that you can check your solution
is reasonable.

1 A box of apples contains 25 apples. How many apples are there in
9 boxes?

2 Sam's school is 29 km from his home.

   a What is the distance to school and back?

   b How far does Sam travel when he goes to school and back
5 times a week?

   c How far does Sam travel in two weeks?

3 Salma orders 57 sets of pens. Each set contains one red,
one blue and one black pen. How many pens are in
57 sets?

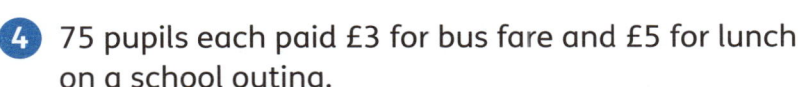

4 75 pupils each paid £3 for bus fare and £5 for lunch
on a school outing.

   a How much money did the pupils pay altogether?

   b What is double this amount?

5 At a school assembly there are 7 rows of 45 chairs and 10 rows of
55 chairs. How many chairs are there altogether?

6 Jaime's school is 483 m from home. He walks to school and back
each day.

   a How far does he walk to school and back in a day?

   b How many metres does he walk to school and back in
five days?

7 On average, 397 cars pass through
a toll station each hour. How many
cars pass through the toll
station in 8 hours?

1. Lisa is making fruit cakes. She needs 485 grams of chopped mixed fruit for one cake.

   a. Lisa plans to make 8 cakes. How much mixed fruit will she need?

   b. The mixed fruit is sold in two different-sized bags:

   Lisa buys enough mixed fruit for 8 cakes, but not enough for 9. What combination of bags could she buy?

2. Each pupil in a group made up a 3-digit by 1-digit multiplication using these number cards:

   5 6 7 8

   Nitha
   $$\begin{array}{r} 876 \\ \times\ \ \ 5 \\ \hline \end{array}$$

   Leon
   $$\begin{array}{r} 758 \\ \times\ \ \ 6 \\ \hline \end{array}$$

   Mari
   $$\begin{array}{r} 657 \\ \times\ \ \ 8 \\ \hline \end{array}$$

   Sulin
   $$\begin{array}{r} 568 \\ \times\ \ \ 7 \\ \hline \end{array}$$

   a. Nitha adds her product to her partner's product and gets a total of 9636. Who is her partner?

   b. What is the sum of the products of the other two pupils?

   c. What is the greatest possible product using these four cards in this way?

Work with line graphs

Line graphs

Think and share

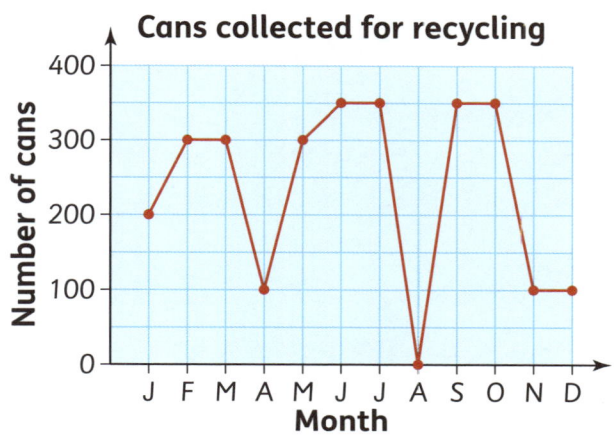

Cans collected for recycling

We often use a **line graph** to show how things change over a period of time.

- What data does this graph show?
- Why is this type of graph is called a line graph?
- Where is time shown on this graph?
- What does it mean when a line goes up from one month to another?
- What does it mean if the line goes down?

1 Study this line graph and then answer the questions.

a What does this graph show?

b What is shown on the horizontal axis? What units are used?

c What is shown on the vertical axis? What units are used?

d How tall was the plant on Sunday?

e How many millimetres did the plant grow from Sunday to Tuesday?

f When did the plant reach a height of 2.5 cm?

g When do you think the plant reached a height of 20 mm? Why?

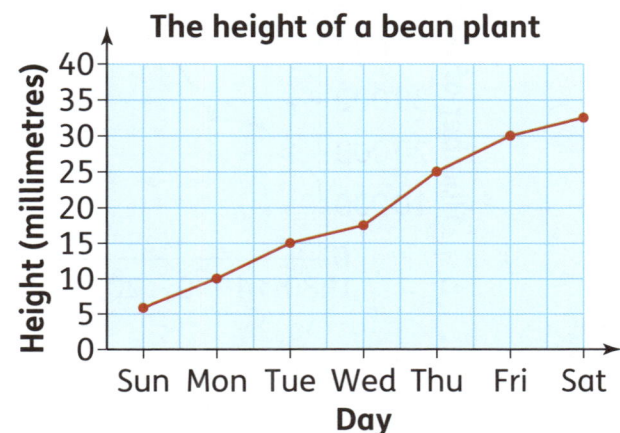

The height of a bean plant

Draw and interpret line graphs

1 This table gives data about the number of people in a hall during an art exhibition.

Time	3 p.m.	4 p.m.	5 p.m.	6 p.m.	7 p.m.	8 p.m.	9 p.m.
Number of people in hall	2	12	56	68	100	100	0

a Draw a grid like the one to the right and use it to plot a line graph showing the data in the table. Make sure you include a title.

b Write a few sentences describing what the graph tells you.

c What time do you think the exhibition opened? Why?

d At what time do you think the exhibition closed? Why?

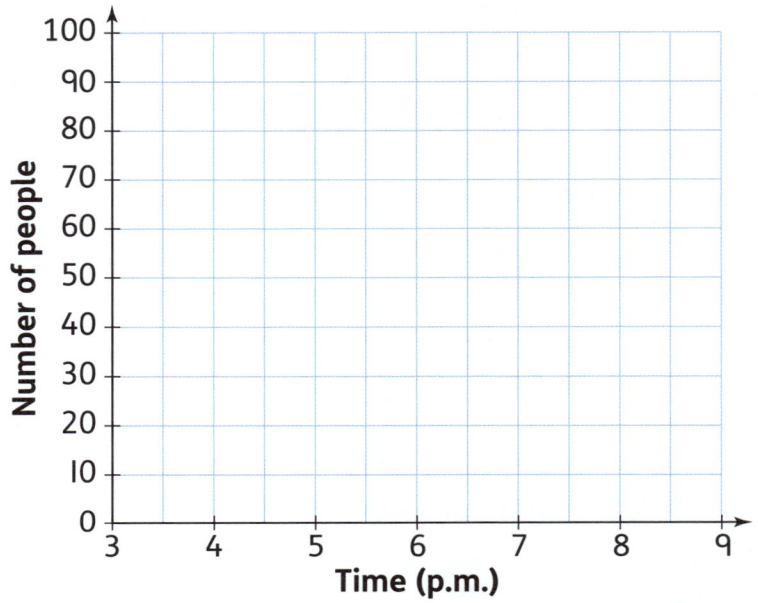

2 a Study the below line graph, which shows the number of puffins on an island in the North Sea from 1980 to 2020.

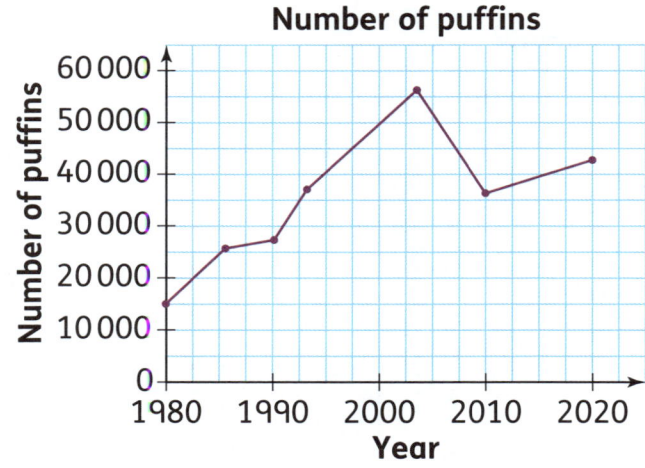

lesson continues

b Work in pairs. Read these statements about the line graph in part a on page 128. Say whether each statement is true or not and give evidence from the graph to support your answers.

The number of puffins was greatest in 2004. There were over 55 000 puffins at this time.

Puffin numbers have decreased steadily since 2005.

From 1990 to 2000, the numbers of puffins increased faster than it did between 1980 and 1990.

Between 2000 and 2010 there was an increase in the number of puffins.

Overall, the number of puffins has increased since 1980.

There were almost 44 000 puffins on the island in 2019.

There was a big increase in the number of puffins from 1985 to 1990.

 Problem solving

3 Here are the answers to some questions about the graph. Write a suitable question for each answer.

   a In 1995   **b** In 2010 and 1993

   c Between 2003 and 2010   **d** Approximately 15 000 birds

   e A difference of approximately 10 000 birds

▶ *Workbook page 81*

Use line graphs to answer questions

1 This line graph shows how the depth of water in a tidal pool changed over a period of six hours.

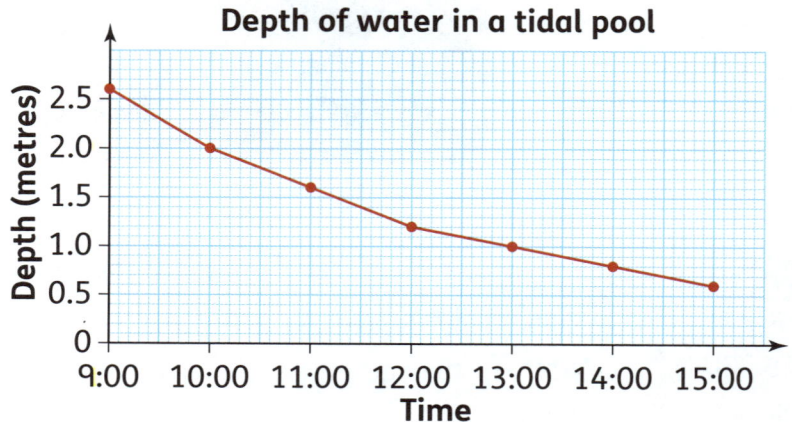

Depth of water in a tidal pool

a How do you know the water level was going down between 9:00 and 15:00?

b Estimate the depth of the water at:

   i 11:30   **ii** 13:30

c Estimate the time when the depth was 1.5 m.

d At 15:00 the tide turned and began to come in. What do you think will happen on the graph after 15:00? Give a reason for your answer.

2 This graph shows the number of students who were absent from school each day last week.

a Describe how the number of students absent changed over the week.

b Explain why the in-between values on this graph have no meaning.

c Which type of graph would be better to show this data?

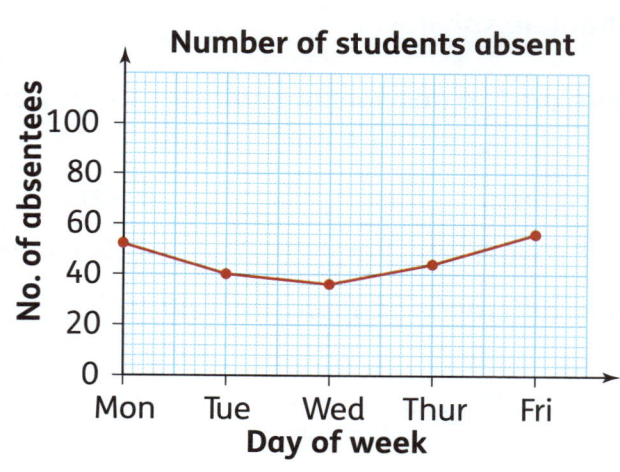

Number of students absent

➡️ *Workbook page 82 and 83*

Patterns

Investigate patterns and rules

💭 Think and share

A sequence is a number pattern that follows a rule to get from one **term** to the next. Can you work out the rule for each sequence?

The **term-to-term rule** tells you how to get from one term to the next.

Use the term-to-term rules you found to work out the next two terms in each sequence.

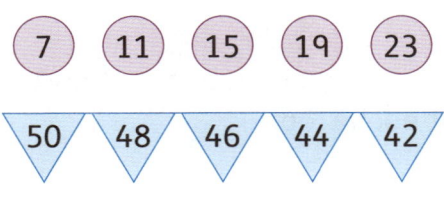

1. What is the term-to-term rule for each of these sequences?

   a 50, 45, 40, 35, 30   b 2, 8, 14, 20, 26

   c 0, −2, −4, −6, −8   d 100, 200, 300, 400, 500

   e 6, 12, 18, 24, 30   f −24, −12, 0, 12, 24

2. Work out the term-to-term rule used in each sequence. Write the next three terms.

   a 2, 7, 12, 17, 22   b 30, 21, 12, 3, −6

   c −3, −6, −9, −12, −15   d 1, 4, 16, 64, 256

3. What are the missing numbers in each sequence?

   a 2, 8, __, 20, __, __   b −24, __, −8, 0, __, __

   c 36, __, 14, 3, __, __   d 1600, 800, __, __, __, 50

   e 7, 14, __, 56, __, __

💡 Problem solving

4. The last term in a sequence that has six terms is 36. What could the sequence be? Give at least three options.

5. A number sequence starts with −12 and ends with 12. Write down three possible sets of numbers for this sequence.

➡ *Workbook page 84*

Number patterns

This tree diagram shows how cells split into two.

Start   **1st split**   **2nd split**

This number sequence shows the number of cells at each step.

1   2   4

1   **a**   Use counters to model how the cells continue to split.

    b   Draw a tree diagram to show how many cells there are each time they split. Continue up to the 6th split.

    c   How many cells are there after 5 splits?

    d   How many cells will there be after 7 splits?

    e   Write a number sequence to show the number of cells up to the 10th split.

    f   What is the rule for this number sequence?

2   Pauline used 1-cm square tiles to make picture frames.

You could model these using square tiles.
Or you could draw them on 1-cm square paper.

She wrote down the number of tiles in each one.
Her number sequence was 10, 14, 18.
She then measured the perimeter of each frame with a ruler.
The perimeters made a different sequence: 14 cm, 18 cm, 22 cm.

    a   Continue her pattern to make the next three picture frames.

    b   Write down the number sequence for the tiles in the frames.

    c   What is the rule for the number of tiles in each frame?

    d   Write down the number sequence for the perimeters of the frames.

    e   What is the rule for the perimeter of each frame?

Square numbers

We can draw dots to represent numbers.

1   2   3   4

If you can arrange the dots in the shape of a square, then the number of dots in the square is a **square number**.

We can draw 9 dots:

Each side of the square is 3 dots long.

$3 \times 3 = 9$

We say 3 squared is 9.

1 Is each number a square number? Draw dot diagrams to help you.

    a 25   **b** 81   **c** 40   **d** 35   **e** 48

2 Mara wrote down the first ten square numbers. She was trying to find a rule for this sequence.

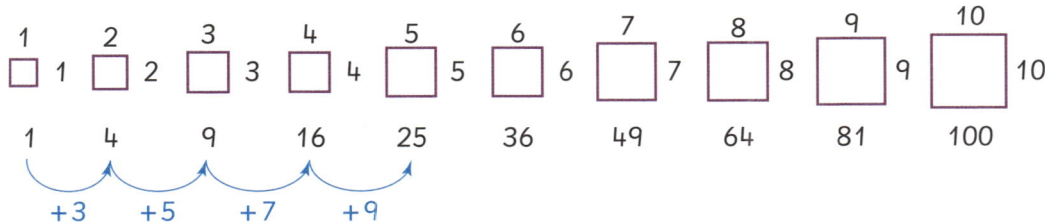

    a What pattern can you see in her work?

    b Discuss how you can use this pattern to make a term-to-term rule for this sequence.

Problem solving

3 Read what three pupils said. Are they correct or not?

a
> 2 squared is 4, so 20 squared must be 40.

b
> If you add the first six odd numbers, you will get the sixth square number.

c
> If you square an odd number, the answer will always be odd.

Workbook page 85

Division

Divide using facts you know

💭 Think and share

There are 48 crayons in this box.

- If you shared the crayons equally among 4 children, how many would each child get?

- A teacher makes groups of 6 crayons. How many groups can she make?

- If you took half the crayons, how many would be left?

- What happens if you divide the crayons into 5 groups?

Some numbers do not divide exactly into others.

For example, what is 48 ÷ 5?

Think, what multiple of 5 is close to 48?

$5 \times 9 = 45$,   So, $45 \div 5 = 9$ $48 - 45 = 3$

So, $48 \div 5 = 9$ with 3 left over.

The amount left over is called a **remainder**. We can write this 9 r3.

1 For each jar of beads, work out how many groups of 5 you can make. Write how many beads are left over.

a	b	c	d	e	f	g
19	42	51	67	79	83	99

💡 Problem solving

Can you use a diagram to help you?

2 When you multiply two numbers you get 54. When you divide 54 by one of the numbers you get 6.
What are the two numbers?

More division

There are different ways to record your work when you divide.

41 ÷ 4

$$
\begin{array}{r}
41 \\
- 40 \quad\quad 10 \times 4 \\
\hline
1
\end{array}
$$

41 ÷ 4 = 10 remainder 1

48 ÷ 3

$$
\begin{array}{r}
16 \\
3\overline{)48} \\
30 \quad\quad 10 \times 3 \\
\hline
18 \\
18 \quad\quad 6 \times 3 \\
\hline
0 \quad\quad \text{16 groups of 3}
\end{array}
$$

No remainder

48 ÷ 3 = 16

1 Divide by 5. Say how many beads will be left over.

a   47

b   25

c   89

2 Divide. Use the facts you know to help you and show your working out.

a 20 ÷ 3   b 42 ÷ 10   c 40 ÷ 6   d 43 ÷ 5

e 71 ÷ 10   f 49 ÷ 6   g 58 ÷ 8   h 48 ÷ 9

Draw a bar model to show the problem.

Problem solving

3 A group of 29 students want to get into two lines of the same length to play a game. Is this possible? Explain your answer.

4 Sindi has to fill a 60-litre tank with water. She has an 8-litre bucket. How many bucketfuls will she need to fill the tank?

➡ *Workbook page 86*

Divide by 10 and 100

Division is the inverse of multiplication. So, look what happens when you divide by 10 or 100.

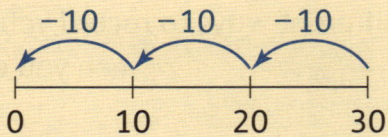

$30 \div 10 = 3$

Thousands	Hundreds	Tens	Ones
		3	0
			3

$1200 \div 100 = 12$

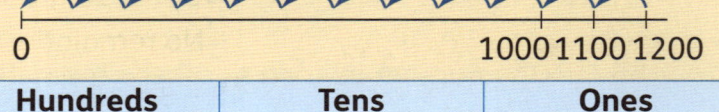

Thousands	Hundreds	Tens	Ones
1	2	0	0
		1	2

- When you divide by 10, the digits move one place to the right.
- When you divide by 100, the digits move two places to the right.

1 Try to do these divisions mentally.

- **a** $40 \div 10$
- **b** $80 \div 10$
- **c** $110 \div 10$
- **d** $240 \div 10$
- **e** $350 \div 10$
- **f** $490 \div 10$
- **g** $520 \div 10$
- **h** $670 \div 10$
- **i** $990 \div 10$

2 Try to do these divisions mentally.

- **a** $400 \div 10$
- **b** $400 \div 100$
- **c** $4000 \div 100$
- **d** $800 \div 10$
- **e** $800 \div 100$
- **f** $8000 \div 100$

Problem solving

3 Sal puts 450 books onto 10 shelves. She puts an equal number of books onto each shelf. How many books are there on each shelf?

4 There are 57 pupils in a year group. The pupils are put into groups of 10. How many pupils will be left over?

▶ *Workbook page 87*

Divide or multiply?

Remember that division is the inverse of multiplication.

If you know that $5 \times 9 = 45$, then you can work out that $45 \div 9 = 5$ and $45 \div 5 = 9$.

1 For each multiplication, write two related division facts.

   a $14 \times 9 = 126$ **b** $15 \times 6 = 90$ **c** $14 \times 8 = 112$

   d $19 \times 3 = 57$ **e** $18 \times 7 = 126$ **f** $9 \times 13 = 117$

   g $8 \times 11 = 88$ **h** $16 \times 5 = 80$ **i** $4 \times 17 = 68$

2 Work out the missing number in each number sentence.

   a $4 \times \boxed{} = 48$ **b** $\boxed{} \times 2 = 24$ **c** $\boxed{} \times 3 = 36$

   d $3 \times \boxed{} = 39$ **e** $8 \times \boxed{} = 48$ **f** $5 \times \boxed{} = 85$

 Problem solving

> Read each question and decide what operation to use.

3 Joe buys big bags of sweets to sell.
He packs the sweets into smaller packets to sell them.

   a For each big bag below, work out how many packets Joe can fill.

     Write down how many sweets will be left over.

SWEETS 89	SWEETS 89	SWEETS 89	SWEETS 89	SWEETS 89
6 in a packet	8 in a packet	10 in a packet	7 in a packet	4 in a packet

   b How many sweets are left over altogether?

   c Joe sells all the leftover single sweets for 5p each.
     How much money does he get?

➡ *Workbook page 88*

1 Look at this picture of an ant.

   a The ant in the picture is 20 mm long.

     This is 4 times longer than the real ant.

     How long is the real ant?

   b In a group of ants, there are 114 legs.

     How many ants are there in the group?

   c How many legs would there be if there were 100 ants?

An ant has 6 legs.

2 Maria makes beaded bangles.
She needs 5 red beads, 10 blue beads and 8 yellow beads for each bangle.

   a She has 420 blue beads.
How many bangles can she make?

   b She has 97 red beads.
How many bangles can she make?

     Will she have any red beads left over?

   c How many yellow beads will she need to make 32 bangles?

3 A bee has to visit about 4400 flowers to get enough nectar to make 10 grams of honey. How many flowers does the bee need to visit to get enough nectar for 1 gram of honey?

4 Aunty Zelma needs these ingredients to make 12 buns.

1 cup flour	4 teaspoons baking powder
$\frac{3}{4}$ cup milk	$\frac{1}{2}$ cup sugar
3 eggs	80 ml of melted butter

Rewrite the ingredient list to show the amounts she needs to make:

   a 24 buns   **b** 36 buns.

5 This spider is three times the size of the real spider.
How could you work out the length of the real spider?

Mixed practice 3

1 Calculate.

    a 27 × 5   **b** 19 × 3   **c** 42 × 6

    d 36 × 8   **e** 27 × 9   **f** 18 × 8

2 Computers are £429 each. How much does it cost to buy three computers?

3 Divide.

    a 790 ÷ 10   **b** 840 ÷ 10   **c** 3200 ÷ 100

    d 56 ÷ 5   **e** 43 ÷ 4   **f** 47 ÷ 6

4 Say whether each number sentence is true or false.

    a 430 × 20 = (400 × 20) + (30 × 10 × 2)

    b 430 × 20 = (430 × 10) + (430 × 10)

    c 430 × 20 = 430 × 10 × 10

    d 430 × 20 = 200 × 43

5 Use the facts that you know to do these multiplications mentally.

    a 50 × 3   **b** 70 × 4   **c** 90 × 8

    d 8 × 60   **e** 7 × 50   **f** 40 × 20

6 Six points have been plotted on the grid.

    a What are the coordinates of points A, C and F?

    b Which point is at (5, 2)?

    c What polygon will you get if you join these points in order from A to F and then joined F to A?

    d What polygon would have vertices at A, B, E and F?

    e Where could you add points G and H to make ABCDEFGH, a symmetrical octagon?

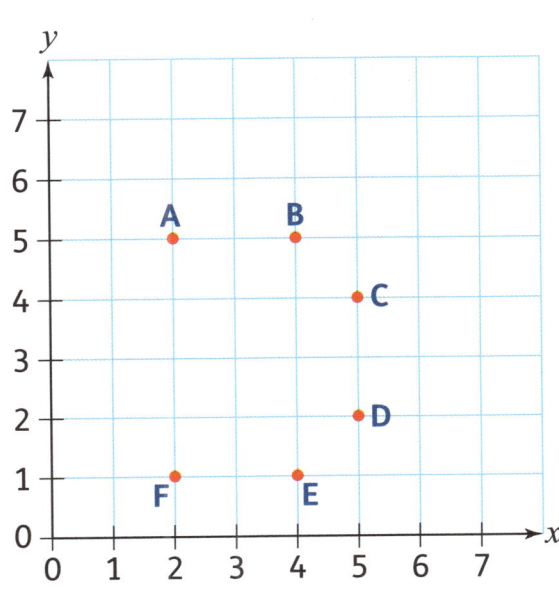

f A triangle is drawn on the grid. It has points A, F and E as its vertices. What type of triangle is this?

g Triangle AFE is translated 2 squares up and 2 squares left. Write the coordinates of its vertices after this translation.

7 A teacher wants to give each of the 34 pupils in her class 5 markers. There are 9 markers in a pack. Each pack costs £4.

   a How many markers does she need in total?

   b How many packs of markers will she need to buy?

   c How much will it cost to buy these packs of markers?

   d Will she have any markers left over? How many?

8 Look at this number sequence:

10, 25, 40, 55, 70, ...

   a What is the term-to-term rule?

   b What are the next three terms in the sequence?

   c Which of these numbers could *not* be terms in this sequence? Explain why.

320	325	355

9 Study the graph carefully. Answer the questions.

   a What does this graph show?

   b How much water was in the barrel at 8:00?

   c When was the water level 20 litres?

   d What happened to the water level from 10:00 to 11:00?

   e How much water was in the barrel at 11:00?

   f When did the water stop flowing out of the barrel? How do you know this?

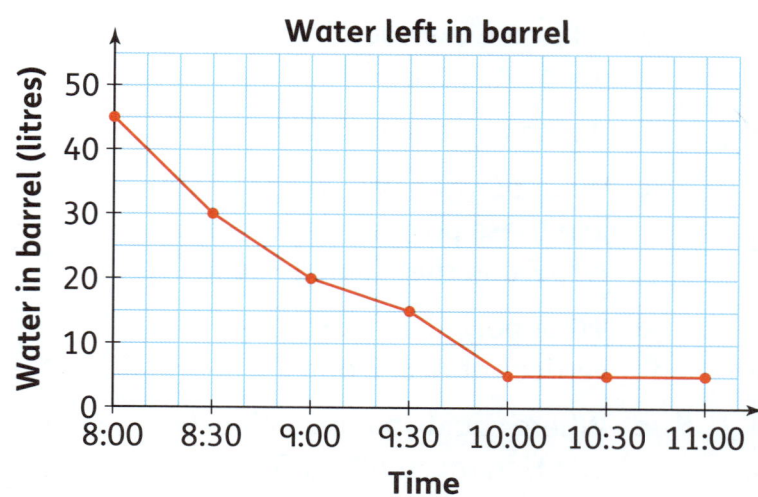

Water left in barrel

Glossary

24-hour notation – a way of recording times using the 24-hour clock, using the numbers 0 to 23 to show the hours in a day. 01:00 is 1 o'clock in the morning (or 1:00 a.m.) and 13:00 is 1 o'clock in the afternoon (or 1:00 p.m.)

A

acute angle – an angle measuring less than 90 degrees

acute-angled triangle – has all its angles less than a right angle

a.m. – stands for *ante meridiem*, which is Latin for 'before noon'. Use to show times before noon, but after 12 midnight

analogue clock – a clock that shows the time using hands that point to numbers arranged in a circle

angle – formed when two lines meet at a point (vertex); the size of an angle is the amount of turn from one of the lines to the other

area – the amount of space covered by or inside a 2D shape, measured in square units such as cm^2

array – objects or pictures arranged in equal rows and columns

axis – two number lines, vertical and horizontal, on a graph or coordinate grid; on a graph, the horizontal axis can be labelled with words instead of numbers

B

bar chart – a graph that uses horizontal or vertical bars to show different categories of data; the lengths of the bars show frequencies

bar model(s) – a diagram used to represent a problem

C

capacity – the amount of liquid a container can hold, measured in units such as millimetres (ml) and litres (ℓ)

Carroll diagram – a table for sorting items according to two sets of categories

centimetre (cm) – a metric unit of length; your thumb is about 1 cm wide

classify – to put things into groups with other things of a similar type

coordinates – two numbers that describe a position on a grid, map, chart or graph

coordinate grid – a grid formed by a pair of numbered axes; the x-axis is horizontal and the y-axis is vertical

counting sequence – a pattern of numbers that go up or down in jumps of the same size

currency – the system or type of money that a country uses

D

decimal(s) – a decimal fraction; with digits to the right of the decimal point, representing tenths, hundredths and so on

degrees – the unit used for measuring angles, for example, 90°

degrees Celsius (°C) – the unit used for measuring temperature, for example 40 °C

difference – you can find the difference between two amounts by subtracting; the difference between 4 and 6 is 2

digit – a symbol used to make a number. 0, 1, 2, 3, 4, 5, 6, 7, 8 and 9 are digits

digital – a way of showing the time using numbers and dots, 07:00 is the digital time for 7 o'clock

E

equilateral triangle – a regular three-sided polygon; a triangle with three sides the same length and three equal angles

equivalent fractions – fractions with the same value; for example, $\frac{1}{2}$ is equivalent to $\frac{4}{8}$

expanded form – a way of writing a number to show the value of each digit

estimate – a sensible guess; an amount calculated using rounded numbers

F

fact family – related facts using the same set of numbers, for example,
$2 \times 3 = 6, 3 \times 2 = 6, 6 \div 3 = 2, 6 \div 2 = 3$

factor(s) – a whole number that divides exactly into another number without leaving any remainder, for example, 5 is a factor of 10 because $10 \div 5 = 2$

fraction – an equal part of a whole object or group; one half, one quarter and one third are all fractions

frequency – the number of times it occurs

frequency table – a table showing how many times a value or an event occurs

G

gram (g) – a metric unit of mass used for light objects; a gram is about the mass of two paperclips; 1000 grams is a kilogram

grid – a pattern of lines, or lines of dots that cross, usually forming squares

H

height – how tall something is

horizontal – a horizontal line goes from one side to the other, parallel to the horizon

hundredths – when you share 1 whole into 100 equal parts, each part is one hundredth; 1 hundredth $= \frac{1}{100} = 0.01$

I

improper fraction – a fraction greater than 1 where the numerator is greater than the denominator; for example, $\frac{5}{3}$

inverse operations – undo an operation; addition and subtraction are inverse operations; multiplication and division are inverse operations. For example, $2 + 3 = 5, 5 - 3 = 2$

isosceles triangle – a triangle with 2 sides equal in length and 2 equal angles where these sides meet the third side

K

key – the information that tells you what each picture represents on a pictogram, or what each colour represents on a diagram

kilogram (kg) – a metric unit of mass used for measuring heavy objects; a kilogram is 1000 grams

kilometre (km) – a metric unit of length used for measuring long distances; there are 1000 metres in a kilometre

L

length – how long something is, or the distance between two points, measured in units such as millimetres, centimetres, metres or kilometres

line graph – a graph made of line segments that join points to show patterns or changes in the data

line of symmetry – a line that divides a shape into two halves that are mirror images of each other

litre (ℓ) – a metric unit of capacity; there are 1000 millilitres in a litre

M

mass – the amount of matter or material in an object, measured in units such as kilograms and grams

metre (m) – a metric unit of length; there are 100 cm in 1 metre

millilitre (ml) – a metric unit of measure used for small capacities: there are 1000 ml in 1 litre

millimetre (mm) – a metric unit of length used for short measurements; there are 10 mm in 1 centimetre

minus sign – the symbol that shows a number is less than zero, or a negative number; for example, –3

mixed number – a number with a whole number part and a fraction part

multiple – a multiple of a number is the product of that number and a whole number; 3, 6, 9 and 12 are all multiples of 3

N

negative number – a whole number less than zero, written with a minus sign; for example, –2

net – a 2D shape that folds to make a 3D shape

number line – a line that shows numbers in order

O

obtuse angle – an angle between 90 degrees and 180 degrees

obtuse-angled triangle – has one angle greater than a right angle

order(ed) – arrange by value or size, for example, from smallest to greatest or greatest to smallest

P

partition – to share a number into smaller parts

perimeter – the distance around the outside of a 2D shape

pictogram – a chart that uses pictures to show information

place – position

place value – the value of each digit in a number, for example, in 64 the '6' has a place value of '6 tens'

p.m. – stands for *post meridiem*, which is Latin for 'after noon'. Use to show times after noon, but before 12 midnight.

point(s) – a small dot or cross marked on a coordinate grid or graph

polygon – a 2D shape with straight sides

product – the answer when you multiply two or more numbers together

properties – characteristics or features of a number or shape, for example

R

regular (polygon) – a regular polygon has all its sides and angles equal

remainder – the amount left over after dividing by a number; for example, $10 \div 3 = 3$ remainder 1

right angle – 90° angle, or a quarter turn; four right angles make a full turn

right-angled triangle – has one angle of 90°

Roman numerals – an ancient number system that uses seven letters to represent numbers; the first five Roman numerals are I, II, III, IV, V

round (a number) – write a number with zeroes in the place of some digits; to round a decimal number, write the nearest whole number

rounded number – an estimated value that is close to the real number

S

scalene triangle – a triangle with no equal sides or angles

sequence – a number pattern or shape pattern that follows a rule to get from one term to the next

side – a line segment that joins two vertices in a 2D shape

square centimetre – a unit we use to measure area; the area of a square with sides 1 cm long

square number – the product of a number and itself; for example, $4 \times 4 = 16$, so 16 is a square number

symmetry/symmetrical – can be divided into two identical parts; has one or more lines of symmetry

T

tally – to keep score or count

tally mark – a small mark used to count one object; every fifth mark is drawn across the previous four tally marks 卌

term – each number, shape or object in a sequence or pattern is called a term

term-to-term rule – the rule for generating the next term in a sequence or pattern

translation – a movement in one or more directions

turn – to rotate an object around a centre point; a full turn (360°) takes you back to your starting position

V

value – what something is worth

Venn diagram – uses circles to sort objects, shapes and numbers into sets; where the circles overlap shows things that are in both sets

vertical – a vertical line goes straight up or down, perpendicular to the horizon

vertices (vertex) – the points where the sides of a 2D shape meet, or where the edges of a 3D shape meet at a point

volume – the amount of space taken up by an object; volume of a liquid is measured in litres and millilitres

W

width – the measurement of length from one side of a shape to its opposite side

X

x**-axis** – the horizontal axis on a graph

Y

y**-axis** – the vertical axis on a graph

year – a period of 365 days or 12 months